A LETTER FROM JACK DANGERMOND

This map book is a small sampling of the work being done by GIS professionals around the world. GIS maps are increasingly becoming a key in how we understand, approach, and resolve our world's challenges. With the advent of web GIS, these maps are more accessible and the stories they represent are transforming the behavior of organizations and society itself.

Esri® Map Book Volume 34 illustrates how spatial thinking is helping professionals from many fields make better decisions and then act on them in a new way. GIS is also helping improve collaboration and creating a growing culture of holistic problem solving, one that integrates multiple fields, cultures, and tiers of society.

I congratulate all the contributors in *Esri Map Book Volume 34*, whose maps and stories lay the foundation for a better future.

Warm regards,

Jack Dangermond

CONTENTS

WORKERS COMPENSATION INJURY CONCENTRATIONS BY COUNTY–STATE OF FLORIDA, 2016

Western Valley Consulting
Goleta, California, USA
By Cort Colbert

The treatment of workers compensation injury patients is one of the more complex medical treatment systems in the US. For this project, we wanted to determine the relationship of location-based factors that contribute to workers compensation cases. For example: Are the industries in the patients' area a factor? Is the geographic location of the employer a factor? What body parts are most reported as injured in a given area?

The data developed in this project shows that, on average, there are higher levels of workers compensation injuries vs. job count in counties with greater numbers of manual labor jobs (mining, goods-producing, construction, etc.). There also appears to be higher levels of injuries to the back and legs in these areas, when compared to total job count. This would indicate that the areas that have higher concentrations of manual labor jobs are more at risk for those types of injuries than workers in other areas and industries.

These maps were created to allow us to share this data with our clients (physicians) as well as provide them with location analytics beyond just a spreadsheet. Using these data and maps, we were able to assist some of our healthcare partners to determine the locations in the state of Florida where their medical specialties would be most beneficial within the workers compensation system.

CONTACT
Cort Colbert
cort@westernvalleyconsulting.com

SOFTWARE
ArcGIS® Desktop 10.5 and ArcGIS® Pro 2.1

DATA SOURCES
Florida Division of Workers Compensation,
Workers Compensation Insurance Organizations,
Florida Geographic Data Library

Courtesy of Western Valley Consulting.

Arm Injuries

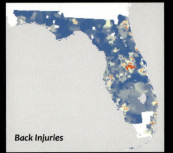
Back Injuries

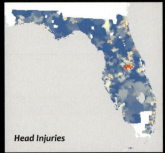
Head Injuries

Leg Injuries

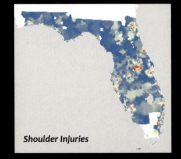
Shoulder Injuries

Torso Injuries

0   15   30   60   90   120
Miles

**Percentage of
Injury Totals / Job Totals**

- 0.38% - 1%
- 1.1% - 2%
- 2.1% - 3%
- 3.1% - 4%
- 4.1% - 5%
- 5.1% - 9%
- 9.1% - 15%
- ⊙ Major Cities

JACKSONVILLE

ORLANDO

CLEARWATER

TAMPA

SAINT PETERSBURG

CAPE CORAL

CORAL SPRINGS

FORT
LAUDERDALE

PEMBROKE
PINES

HOLLYWOOD

HIALEAH

MIAMI

Foot Injuries

Multiple Body Parts

Wrist Injuries

IOT COMPETITION, AMAZON HQ VICINITY

ShareTracker, Inc.
Ashland, Missouri, USA
By Michelle Doyle

With over a 500-mile route near Amazon HQ, ShareTracker discovered a unique Smart Home IoT market base using hotspot analysis. This map shows the concentration and dispersion of over 22,000 smart devices throughout Seattle, Washington. Both Google Home and Amazon Echo share similar coverage among residential neighborhoods near Highlands Park in Renton, Washington.

The collection of sensor data by ShareTracker occurs during the week between the peak hours of 4:00 p.m. and 10:00 p.m., and during the weekend between the hours of 12:00 p.m. and 10:00 p.m. Data collection for smart devices is optimal when kids are home from school and parents are off from work.

ShareTracker utilizes ArcGIS Pro to integrate millions of data points and portray a variety of market trends in IoT, such as home automation, video streaming, and home security. This study highlights just one of the IoT markets across the US. Smart application of IoT data will enable the first adopters to get a share of this growing market.

CONTACT
Michelle Doyle
mdoyle@sharetracker.net

SOFTWARE
ArcGIS Pro 2.0.1

DATA SOURCES
ShareTracker

Courtesy of ShareTracker, Inc.

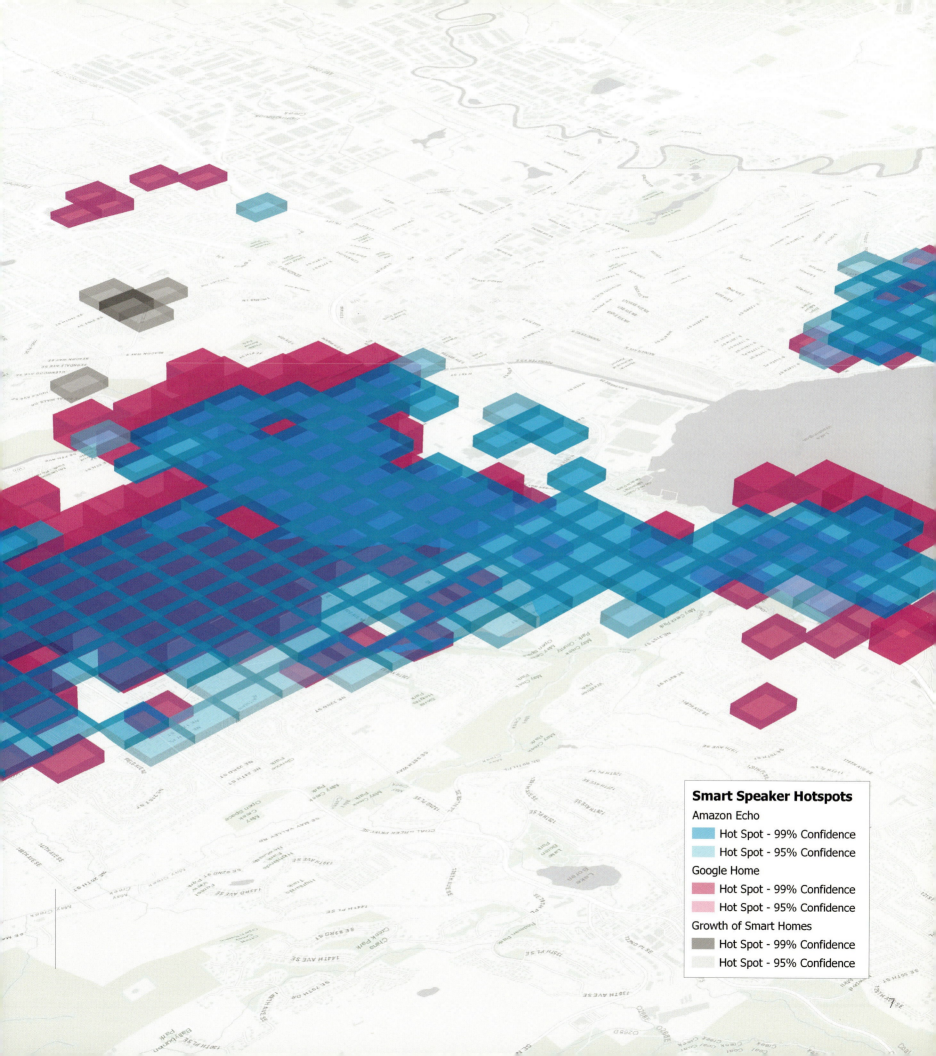

Smart Speaker Hotspots

Amazon Echo

- Hot Spot - 99% Confidence
- Hot Spot - 95% Confidence

Google Home

- Hot Spot - 99% Confidence
- Hot Spot - 95% Confidence

Growth of Smart Homes

- Hot Spot - 99% Confidence
- Hot Spot - 95% Confidence

ECOLOGICAL ATLAS OF THE BERING, CHUKCHI, AND BEAUFORT SEAS

Audubon Alaska
Anchorage, Alaska, USA
By Melanie A. Smith, Max S. Goldman, Erika J. Knight, and
Jon J. Warrenchuk (editors)

The goal of the Ecological Atlas of the Bering, Chukchi, and Beaufort Seas was to create a comprehensive, trans-boundary atlas that represented the current state of knowledge on subjects ranging from physical oceanography to species ecology to human uses.

The Ecological Atlas is organized into six topic areas that build, layer by layer, the ecological foundation of these three seas. Chapter 2 (Physical Setting) explores various climatic attributes and the abiotic processes that perpetuate them. Chapter 3 (Biological Setting) introduces the lower trophic food web. Chapter 4 (Fishes) describes a range of prominent pelagic and demersal fish species. Chapter 5 (Birds) highlights a long list of seabirds and waterbirds that regularly use these waters. Chapter 6 (Mammals) maps out regional use by many cetaceans, pinnipeds, and polar bears. Chapter 7 (Human Uses) covers subsistence, conservation, and economic drivers in the region. These six expansive topic areas culminate in Chapter 8 (Conservation Summary), which shares the key themes and management recommendations stemming from this work.

CONTACT
Erika Knight
eknight@audubon.org

SOFTWARE
ArcGIS Desktop 10.3, Adobe Illustrator®

DATA SOURCES
Audubon Alaska (2009) [based on Eicken et al. (2005)]; Audubon Alaska (2016b) [based on Carmack and MacDonald (2002), Eicken et al. (2009), National Oceanic and Atmospheric Administration (1988), and National Snow and Ice Data Center et al. (2006)]; Audubon Alaska (2016c) [based on Fetterer et al. (2016)]; Audubon Alaska (2016d) [based on Fetterer et al. (2016)]; Audubon Alaska et al. (2017); Carmack and MacDonald (2002); Fetterer et al. (2016); National Snow and Ice Data Center and Konig Beatty (2012); Oceana and Kawerak (2014); Satterthwaite-Phillips et al. (2016); Scenarios Network for Alaska and Arctic Planning (2016); Spiridonov et al. (2011); Stringer and Groves (1991), Audubon Alaska (2017) based on exactEarth (2017), Audubon Alaska (2014); Audubon Alaska (2015); Audubon Alaska (2016h) [based on Fetterer et al. (2016)]; Audubon Alaska (2017d) [based on Audubon Alaska (2014) and BirdLife International (2017a)]; Audubon Alaska (2017j) [based on Audubon Alaska (2016a)]; Audubon Alaska (2017k) [based on Petersen et al. (1999) and Sexson et al. (2014)]; Audubon Alaska (2017l) [based on Audubon Alaska (2015), Audubon Alaska (2016a), BirdLife International (2017a), D. Safine (pers. comm.), eBird (2015), Petersen et al. (1999), Sea Duck Joint Venture (2016), and US Fish and Wildlife Service (2016b)]; BirdLife International (2017a); D. Solovyeva (pers. comm.); Sexson et al. (2012); Sexson et al. (2016); Solovyeva and Kokhanova (2017) [based on Arkhipov et al. (2014), Krechmar and Kondratyev (2006), Solovyeva (2011)]; US Fish and Wildlife Service (2016b)

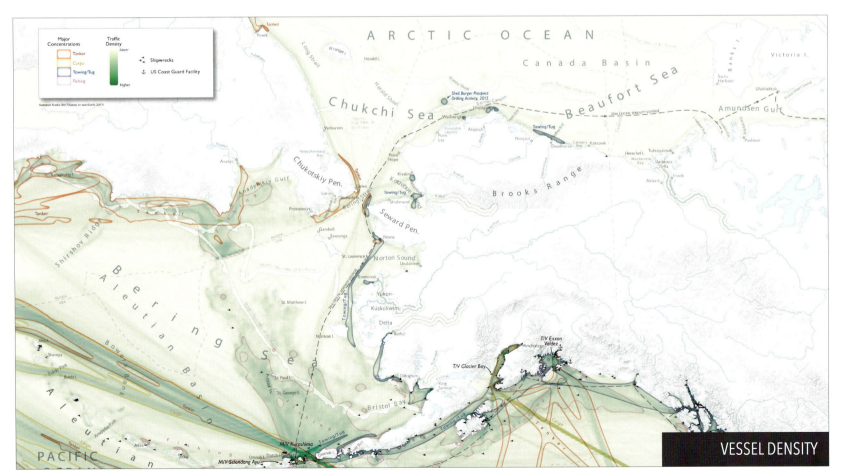

VESSEL DENSITY

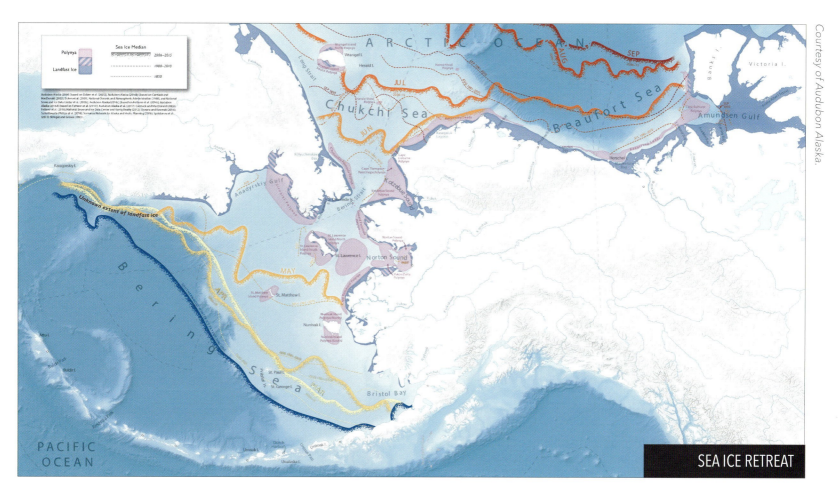

SEA ICE RETREAT

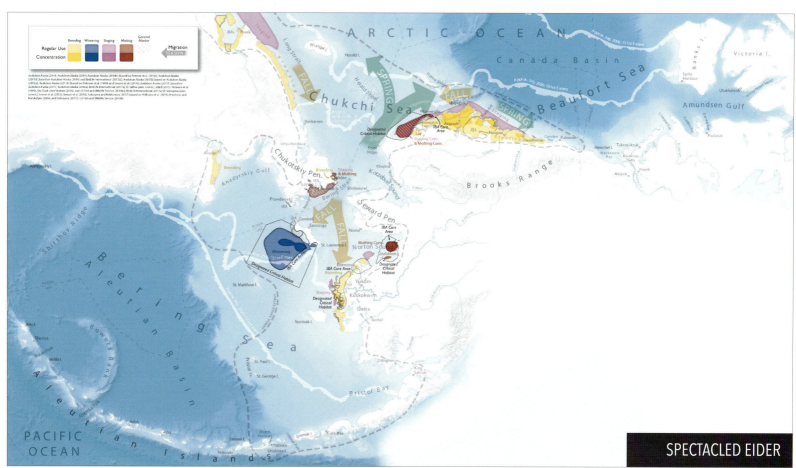

SPECTACLED EIDER

PRELIMINARY SURFICIAL GEOLOGY OF MIAMI COUNTY, KANSAS

Kansas Geological Survey - University of Kansas
Lawrence, Kansas, USA
By John W. Dunham, Anthony L. Layzell, Rolfe D. Mandel,
and K. David Newell

Miami County is located along the Missouri border in eastern Kansas in the Osage Cuestas physiographic region, an area characterized by a series of east-facing ridges (escarpments) alternating with flat to gently rolling plains. The surficial geology map shows bedrock and sediment layers at the surface or immediately under vegetation and soil. This map is the first detailed geologic mapping of the area since 1966.

The youngest surficial deposits in the map area are found in valley landscapes and consist of alluvium (floodplain and terraces) and colluvium deposited within the last 11,000 years. Most surface rocks formed from sediment deposited in ancient seas about 290 to 310 million years ago. The underlying strata of the Osage Cuestas are Pennsylvanian limestones and shales that dip gently to the west and northwest.

The map shows the distribution, rock type, and age of bedrock. It can be used to identify surface and subsurface lithologic units and their stratigraphic relationships, show geologic structures, delineate thicknesses of surficial materials such as alluvium, and determine the features' spatial orientation.

CONTACT
John Dunham
dunham@kgs.ku.edu

SOFTWARE
ArcGIS Desktop 10.5, ArcGIS® Spatial Analyst™

DATA SOURCES
Kansas Department of Transportation, US Geological Survey, Kansas Data Access and Support Center

Courtesy of Kansas Geological Survey-University of Kansas.

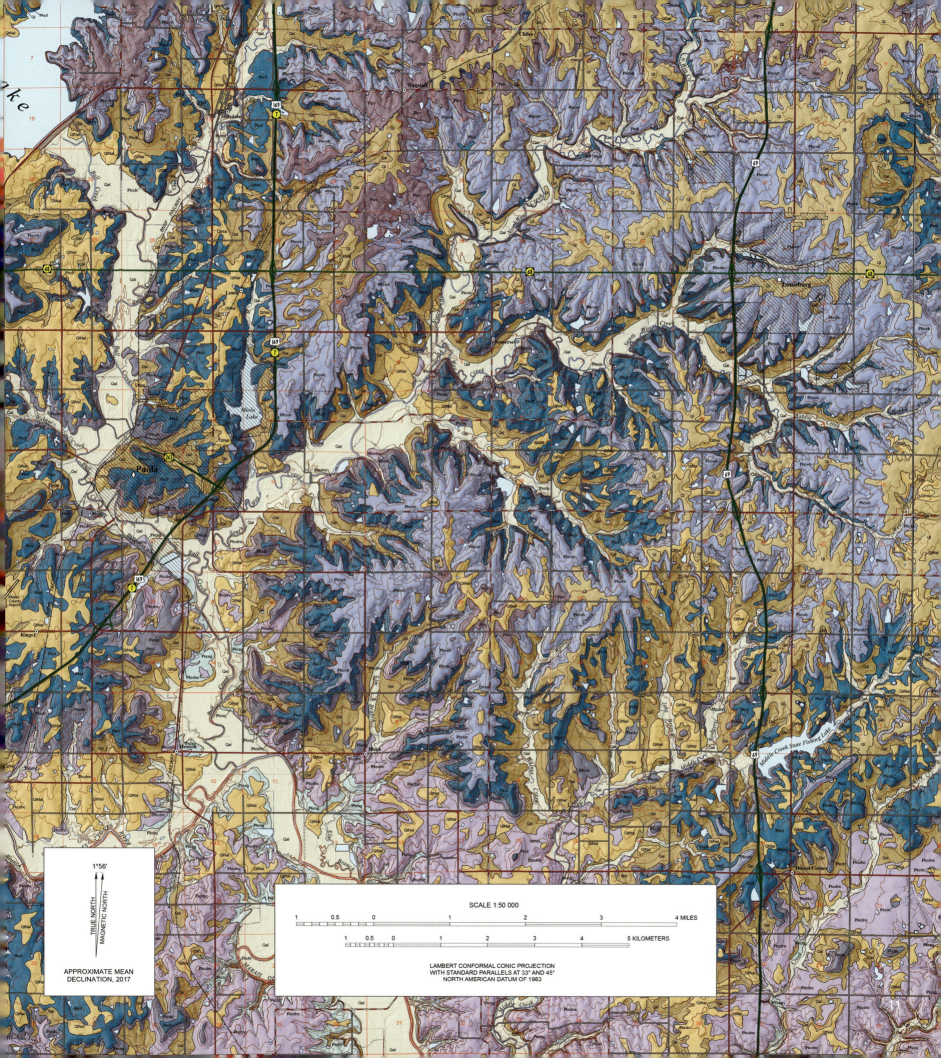

SCALE 1:50 000

1   0.5   0   1   2   3   4 MILES

1   0.5   0   1   2   3   4   5 KILOMETERS

LAMBERT CONFORMAL CONIC PROJECTION
WITH STANDARD PARALLELS AT 33° AND 45°
NORTH AMERICAN DATUM OF 1983

1°56'

TRUE NORTH

MAGNETIC NORTH

APPROXIMATE MEAN
DECLINATION, 2017

Paola

Hillsdale

Louisburg

Rinner

Hensen

Drexel Corner

Middle Creek State Fishing Lake

GRAND TETON NATIONAL PARK– JENNY LAKE TRAILS

David B. Sherman
Austin, Texas, USA
By David B. Sherman

The mountains of Grand Teton National Park are well known for their topographic prominence, with Grand Teton rising some 7,000 feet above the valley floor only four miles to the east. Mapping the contours of such prominent mountains is a challenge, and many published maps are printed at a scale too small for accurate route planning and orienteering. This large-scale map focuses on trails in the vicinity of Jenny Lake, a popular hiking area in the park.

The complex topography of the Teton Range also makes it ideally suited for experimentation with various hillshade techniques. For this map, an open source 3D graphics software called Blender was used to simulate light dispersion and shading over the digital elevation model. The result is a soft hillshade that gives the mountains depth without obscuring the contour detail. The basemap is also the digital elevation model, displayed with a desaturated version of the Esri elevation color scheme in ArcGIS Pro. This gives the user a general sense of the topography without distracting from the trail routes and labels.

CONTACT
David Sherman
david.sherman@clearwayenergy.com

SOFTWARE
ArcGIS Pro 2.1, Blender™ 2.79

DATA SOURCES
US Geological Survey, US National Park Service

Courtesy of Macquarie Capital.

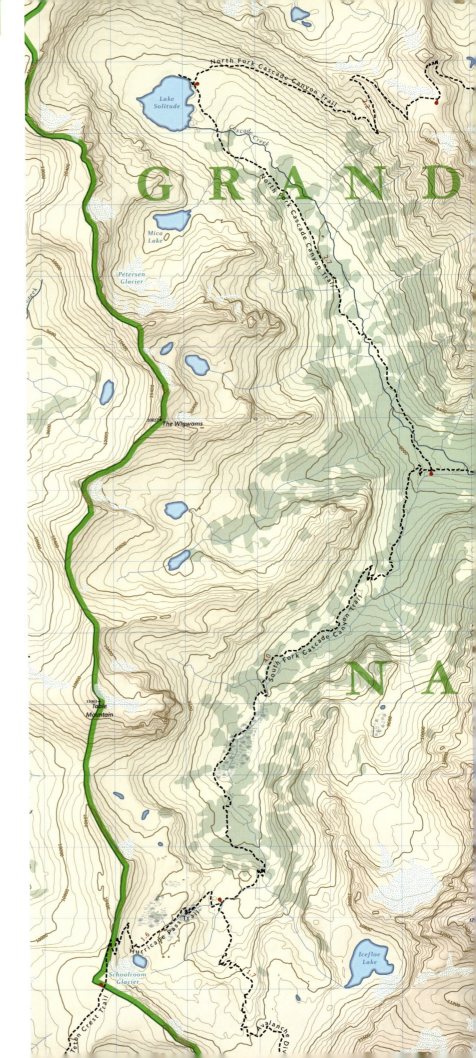

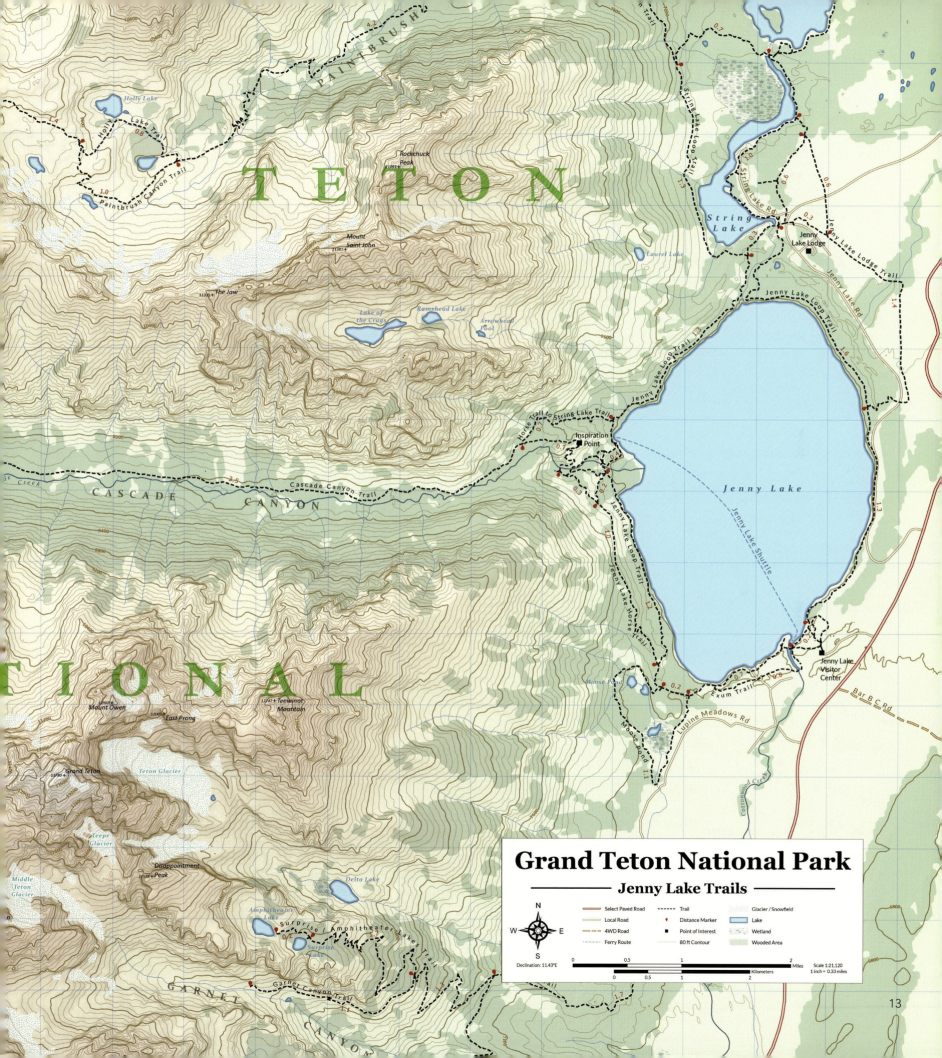

8-BIT ELECTROMAGNETIC ELEVATION EXPERIENCE

Oregon Metro
Portland, Oregon, USA
By Matthew Hampton

Exploring data is just as important as exploring how we structure the data.

Multi-band raster data can be digitally stored in a variety of file formats that typically use one of three common binary schemes: band interleaved by line (BIL), band interleaved by pixel (BIP), and band sequential (BSQ). The information for the computer to understand the file structure is written in the header of the file, but what happens when you change the instructions?

Converting elevation data between a BIL and BIP created this artistic representation of the Pacific Northwest by pivoting the data structure so every 8-bit band of elevation data (256 feet) is interpreted by the computer differently.

Looking at data through different file frameworks can highlight the artistic nature of data and landscape.

CONTACT
Matthew Hampton
matthew.hampton@oregonmetro.gov

SOFTWARE
ArcGIS Desktop

DATA SOURCES
Metro's Regional Land Information System (RLIS)

Courtesy of Matthew Hampton of Oregon Metro.

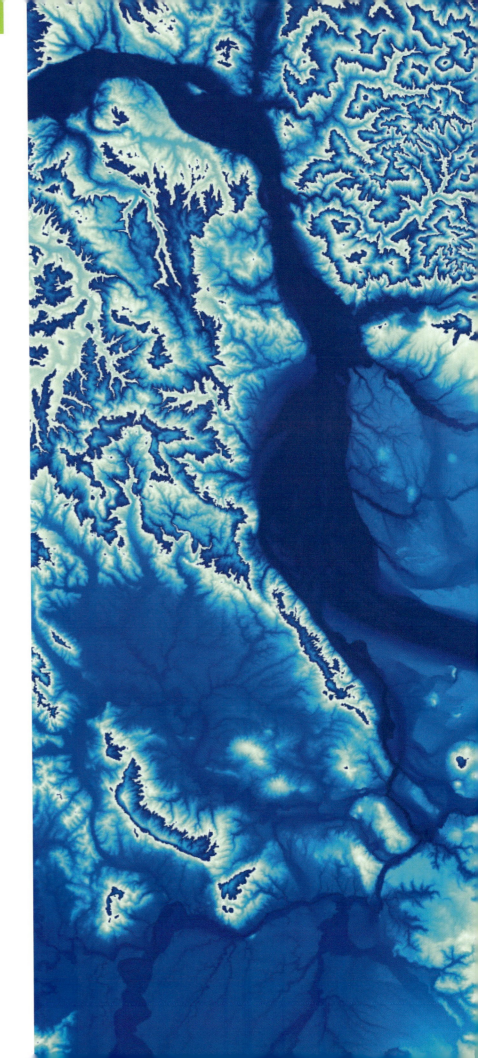

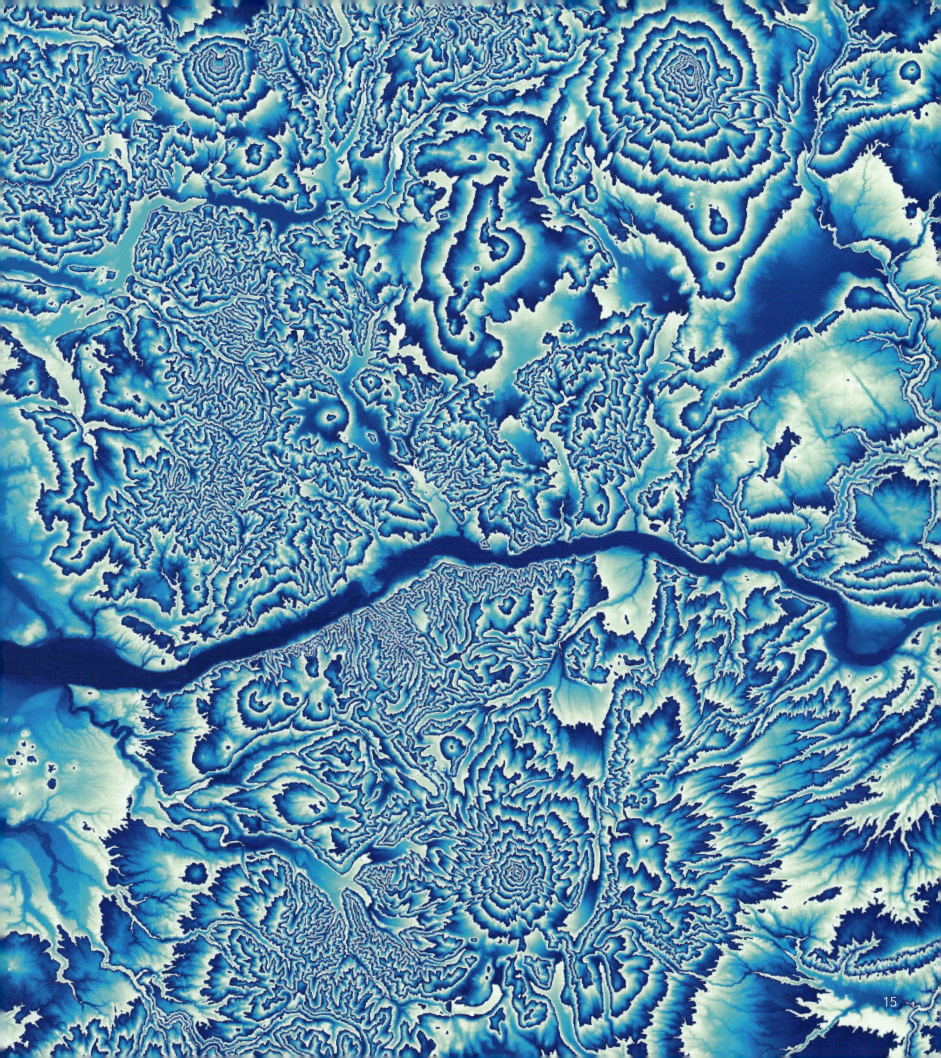

UCSC CAMPUS MAP

UC Santa Cruz
Santa Cruz, California, USA
By Aaron Cole, Parker Welch, and Barry Nickel

UC Santa Cruz is famous for many things, including its contribution to academic research, a long history of questioning authority, and simply the natural beauty of the campus itself. Often cited as one of the most beautiful campuses in the world, UC Santa Cruz's physical setting offers a diverse blend of award-winning architecture and a natural environment of redwood forests, scenic ocean views, and open meadows.

We at the Center for Integrated Spatial Research have long felt that the map representing our campus should be equally as stunning and inspire its readers to visit and explore. In 2017, we completely redesigned the campus map using a combination of lidar, high-resolution aerial imagery, and computer-aided design asset data converted to GIS feature classes. Our goal was to produce a final product with just the right combination of realism and abstraction to make the map as effective for wayfinding as framing on the wall.

CONTACT

Aaron Cole
adcole@ucsc.edu

SOFTWARE

ArcGIS Desktop 10.5.1, Adobe Illustrator Creative Suite® 6, Adobe Photoshop® Creative Suite 6

Courtesy of UC Santa Cruz.

SOCIAL SCIENCES 1

COLLEGE 9

FIREHOUSE

CROWN/MERRILL APARTMENTS

COLLEGE 10

SOCIAL SCIENCES 2

DINING COMMONS

K7SC RADIO STATION

CANTU CENTER

MERRILL COLLEGE

HEAT PLANT

COMMUNICATIONS BUILDING

McLAUGHLIN DRIVE

UNIVERSITY CENTER

DINING COMMONS

PHYSICAL SCIENCES

BIOMEDICAL SCIENCES

RED HILL ROAD

CROWN COLLEGE

MERRILL ROAD

SCIENCE HILL

SINSHEIMER LABS

SCIENCE & ENGINEERING LIBRARY

STUDENT HEALTH CENTER

ALAN CHADWICK GARDEN

McLAUGHLIN DRIVE

COOLIDGE DRIVE

NS 2 ANNEX

INTERDISCIPLINARY SCIENCES

QUARRY AMPHITHEATER

HUMANITIES & SOCIAL SCIENCES DEPT.

THIMANN LABS

NATURAL SCIENCES 2

EARTH & MARINE SCIENCES

BAYTREE BOOKSTORE

HAHN ART FACILITY

STEVENSON COLLEGE

THIMANN LECTURE HALL

CENTER FOR ADAPTIVE OPTICS

CLASSROOM UNIT

STUDENT UNION

COWELL COLLEGE

STEINHART WAY

REDWOOD BUILDING

QUARRY PLAZA

ELOISE PICKARD SMITH GALLERY

EVENT CENTER

KERR HALL

DINING COMMONS

HEULER ROAD

HAHN STUDENT SERVICES

HAGAR DRIVE

MCHENRY LIBRARY

HAHN ROAD

MENTAL THEATER

OPERS OFFICES

MEDIA THEATER

EAST FIELDHOUSE

EAST FIELD

ELENA BASKIN VISUAL ARTS

STAGE

2ND STAGE OFFICES

ACADEMIC RESOURCE CENTER

PHYSICAL EDUCATION, RECREATION, AND SPORTS

PRESS BLDG

GAMELAN STUDIO

DIGITAL ARTS RESEARCH CENTER

OPERS WELLNESS CENTER

MUSIC BUILDING

RECITAL HALL

MUSIC

UNIVERSITY HOUSE

BIKE PATH

LOWER EAST FIELD

EAST REMOTE PARKING

GREAT MEADOW

HAGAR

COOLIDGE DRIVE

17

JUNEAU AREA TRAILS GUIDE

US Department of Agriculture (USDA) Forest Service
Juneau, Alaska, USA
By Robert Francis, Joe Calderwood, Dustin Wittwer, Ed Grossman, Mike Dilger, Brad Orr, Quinn Tracy, Kevin Kolb, Matt Tharp, Michele Gorham, Mike Brown, and Faith Duncan

The Juneau Area Trails Guide depicts the most extensive trail system in southeast Alaska. From accessible trails to historic mining routes and unmaintained trails, Juneau's trail network allows recreationists of all abilities to explore a wide diversity of habitats and scenic attractions.

This version of the Juneau Area Trails Guide takes full advantage of the newly acquired statewide elevation data, an interferometric synthetic-aperture radar (IfSAR)–derived digital elevation model supporting five-meter resolution. IfSAR–derived ground contours illustrate the current orthometric topography, while hillshades from both the ground and surface models are combined to subtly bring out the texture of the landscape. The inset maps provide larger scale detail for key areas of interest and use the actual IfSAR orthorectified radar image (ORI) to provide subtle textural details and serve as a surrogate for hillshade.

CONTACT
Kim Homan
khoman@fs.fed.us

SOFTWARE
ArcGIS Desktop 10.3, Adobe Creative Cloud®, Avenza MAPublisher®

DATA SOURCES
USDA Forest Service corporate data, City and Borough of Juneau GIS, Alaska Department of Natural Resouces GIS, National Oceanic and Atmospheric Administration National Tsunami Hazard Mitigation Program, US Geological Survey-IfSAR, US Geological Survey Global Land Ice Measurements from Space, Randolph Glacier Inventory 6.0

Courtesy of US Department of Agriculture (USDA) Forest Service.

GRAND STAIRCASE-ESCALANTE NATIONAL MONUMENT– RECREATION OPPORTUNITIES

Utah, Bureau of Land Management
Salt Lake City, Utah, USA
By Lynn Roth

The Grand Staircase-Escalante National Monument spans many acres of America's public lands and contains three distinct units, Grand Staircase, Kaiparowits, and Escalante Canyon. The monument's size, resources, and remote character provide extraordinary opportunities for geologists, paleontologists, archeologists, historians, and biologists in scientific research, education, and exploration.

This map shows the existing and authorized recreation opportunities within the Grand Staircase-Escalante National Monument and surrounding areas including campgrounds, hiking trails, information kiosks, points-of-interest, recreation sites, scenic byways and backways, and national historic trails.

Basemap elements used on this map are a combination of shaded relief, digital elevation models, and National Agriculture Imagery Program (NAIP) manipulated to give the map reader an accurate display of the terrain complexities within the Grand Staircase-Escalante National Monument.

CONTACT
Lynn Roth
lroth@blm.gov

SOFTWARE
ArcGIS Desktop 10.4.1, Adobe Illustrator CC, Adobe Photoshop CC

DATA SOURCES
US Geological Survey NAIP, Esri, National Atlas of the United States, Utah Automated Geographic Reference Center, Utah Department of Transportation, BLM Corporate Data

Courtesy of Utah, Bureau of Land Management 2018.

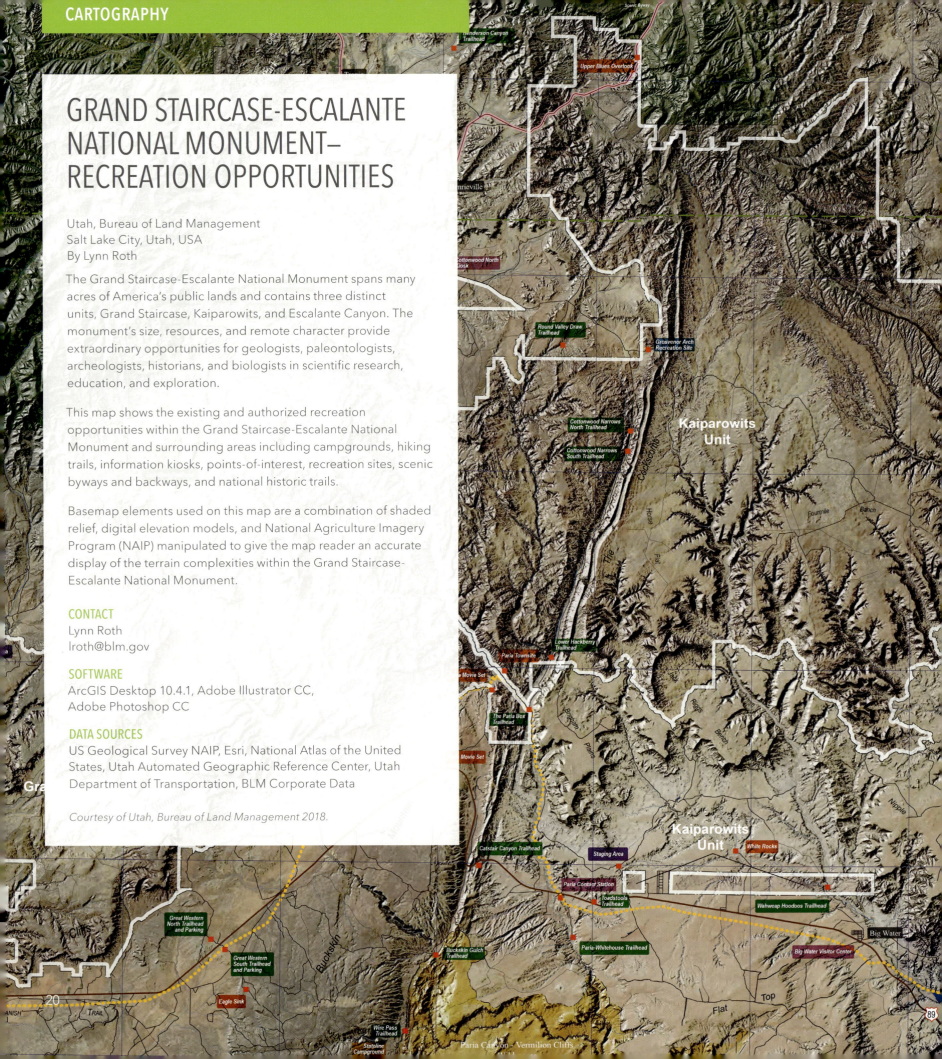

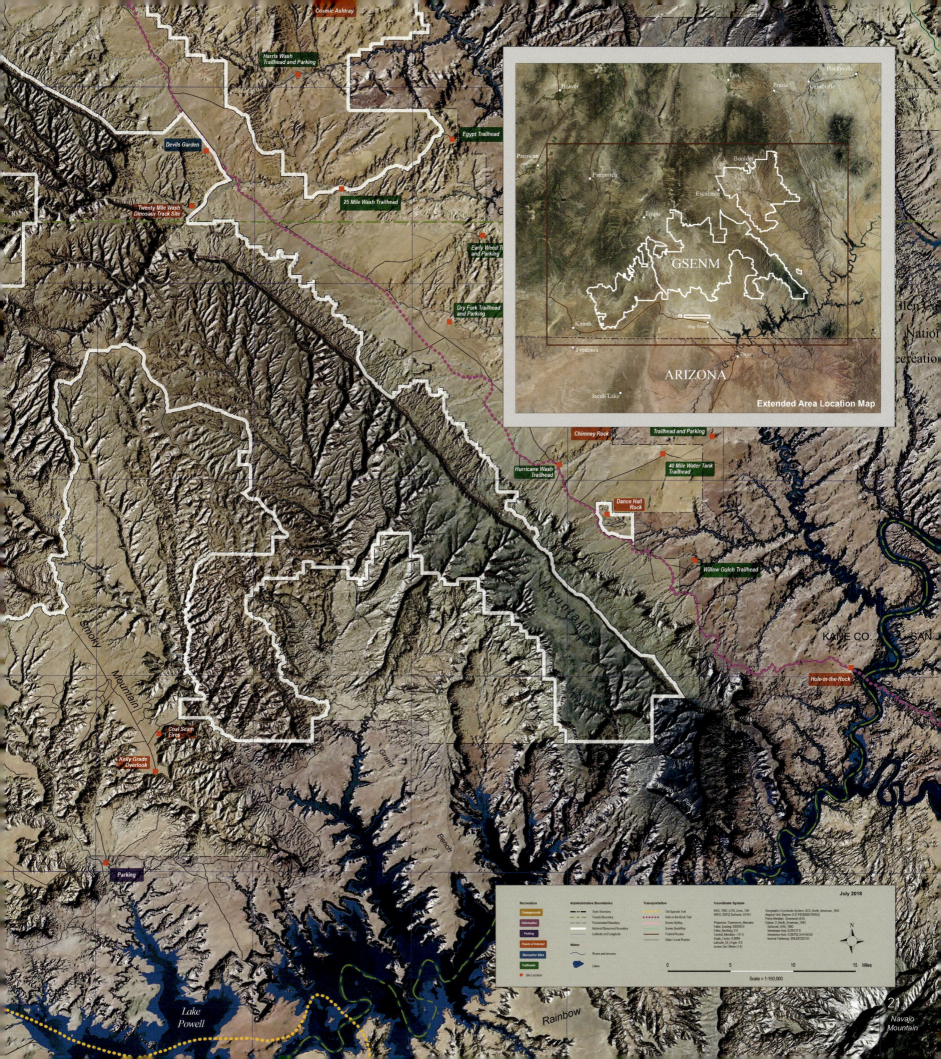

Cosmic Ashtray

Harris Wash
Trailhead and Parking

Egypt Trailhead

Devils Garden

25 Mile Wash Trailhead

Twenty Mile Wash
Dinosaur Track Site

Early Weed T
and Parking

Dry Fork Trailhead
and Parking

Chimney Rock   Trailhead and Parking

40 Mile Water Tank
Trailhead

Hurricane Wash
Trailhead

Dance Hall
Rock

Willow Gulch Trailhead

KANE CO.   SAN

Hole-in-the-Rock

Coal Seam
Fires

Kelly Grade
Overlook

Parking

Lake
Powell

Rainbow

21
Navajo
Mountain

Extended Area Location Map

Heaver   Loa   Fruita   Hanksville

Parowan

Panguitch   Boulder

Escalante

Teasdale

GSENM

Kanab   Big Water

Fredonia   Paria

ARIZONA

Jacob Lake

Nation

recreation

len

SAN

Grand   Smoky   Mountain

Bench

Recreation   Administrative Boundaries   Transportation   Coordinate System   July 2018

Campgrounds   State Boundary   Old Spanish Trail   NAD, 1983, UTM, Zone, 12N
Information   County Boundary   Hole-in-the-Rock Trail   SBEG, 2012 Authority, EPSG
Parking   Proclaimed Boundary   Scenic Byway   Projection, Transverse_Mercator
Points of Interest   National Monument Boundary   Scenic BackWay   False_Easting, 500000.0
Recreation Sites   Latitude and Longitude   Federal Routes   False_Northing, 0.0
Trailheads   State/Local Routes   Central_Meridian, -111.0

Water   Scale_Factor, 0.9996
Site Location   Rivers and streams   Latitude_Of_Origin, 0.0
Lakes   Linear_Unit, Meter (1.0)

0   5   10   15 Miles

Scale = 1:150,000

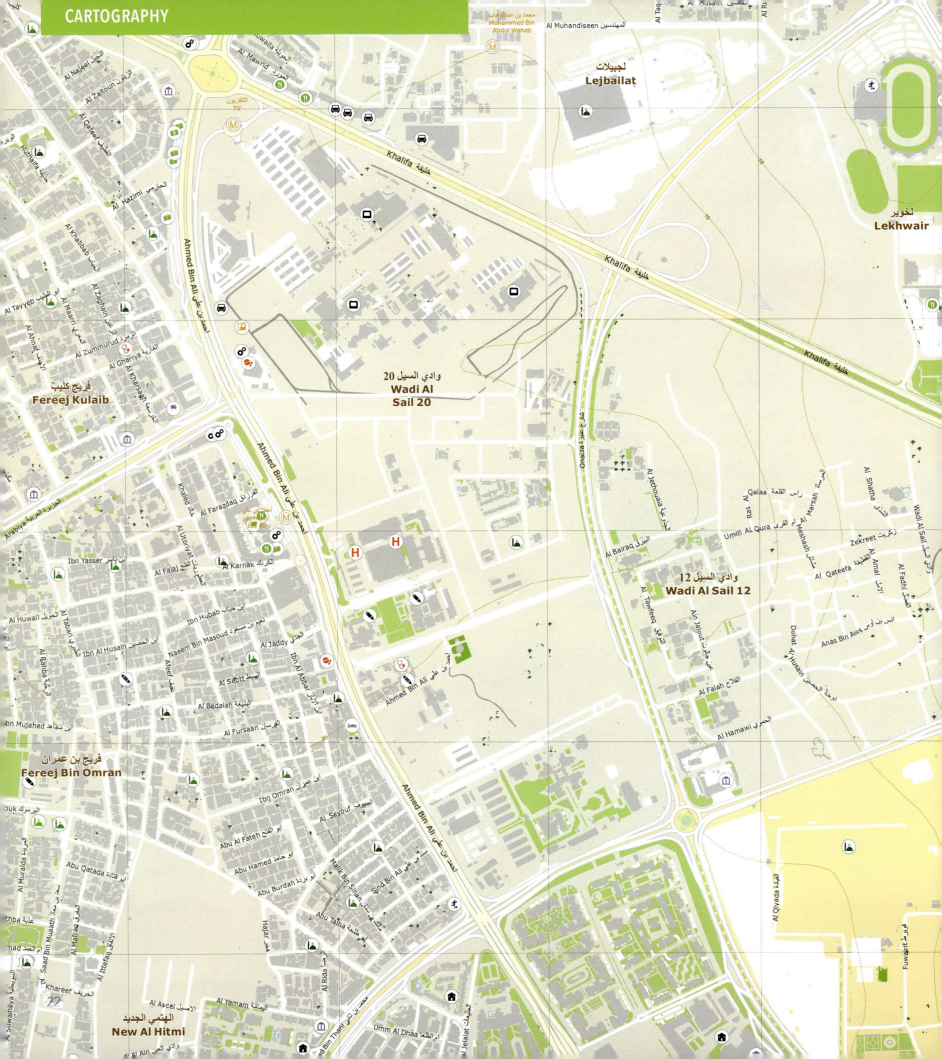

محمد بن عبدالوهاب
Mohammed Bin
Abdul Wahab

المهندسين
Al Muhandiseen

لجبيلات
Lejbailat

لخوير
Lekhwair

Al Najeel النجيل
Al Zaitoun الزيتون
Al Qataei القطاعي
Al Khabbab الخباب
Al Tayyeb الطيب
Al Maari المعري
Al Zaghain الزغين
Al Anai الأناعي

Khalifa خليفة

Al Hazimi الحازمي

Al Zummurud الزمرد
Al Gharriya الغرية
Al Kharsaah الخرساء

Fereej Kulaib فريج كليب

وادي السيل 20
Wadi Al
Sail 20

Khalifa خليفة

Onaiza عنيزة

Ras Al رأس ال
Qalaa القلعة
Al Jethoualia الجذوعلية
Umm Al Qura أم القرى
Al Bairaq البيرق
Al Marsah المرسى
Al Meshash المشاش
Al Shatha الشذا
Wadi Al Sail وادي السيل
Al Amal الأمل
Zekreet زكريت
Al Qateefa القطيفة
Al Fadhi الفضي

وادي السيل 12
Wadi Al Sail 12

Al Tawfeeq التوفيق
Al Jalout عين جالوت
Dohat Al دوحة ال
Husain الحسين
Anas Bin Aws أنس بن أوس

Al Falah الفلاح
Al Hamawi الحموي

Arabiya العربية
Ibn Yasser ابن ياسر
Al Huwail الحويل
Al Tabari الطبري
Al Rahba الرحبة
bn Mujahed ابن مجاهد

Khaled خالد
Al Utorayat العطريات
Al Farazdaq الفرزدق
Al Falaj الفلج
Al Karnak الكرنك
Ibn Hubab ابن حباب
Naeem Bin Masoud نعيم بن مسعود
Ateef عطيف
Al Sebti السبتي
Al Bedaiah البديعة
Al Jaddy الجدي
Ibn Al Abbas ابن العباس
Al Fursaan الفرسان

H
H

Ahmed Bin Ali أحمد بن علي

Al Qiyada القيادة

Fereej Bin Omran فريج بن عمران

uk اليرموك
Al Muraida المريدة
Abu Qatada أبو قتادة
Suwainiya السويدية
bha الشبهة
Saad Bin Muaath سعد بن معاذ

Ibn Omran ابن عمران
Abu Fateh أبو فاتح
Abu Hamed أبو حامد
Al Seyouf السيوف
Malik Bin Sinan مالك بن سنان
Sind Bin Ali سند بن علي
Abu Burdah أبو بردة
Abu Talha أبو طلحة
Hajar هجر

Funwairit فنويرط

Al Khareef الخريف
Al Yamam اليمام
Al Ascel الأصيل
Al Rida الرضا
ed Bin Thani بن ثاني
Umm Al Dhaa أم الظعا
Al Jelaiat الجليعات

New Al Hitmi الهتمي الجديد

TV التلفزيون

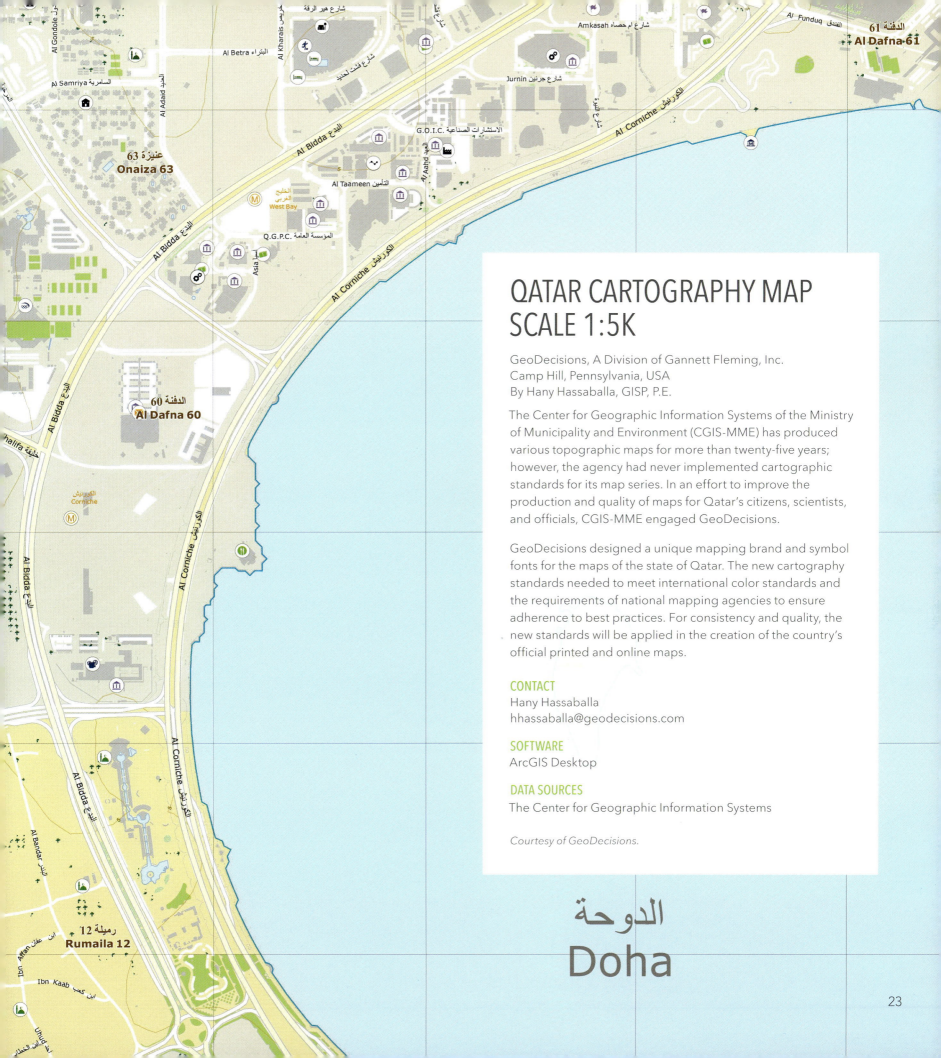

QATAR CARTOGRAPHY MAP
SCALE 1:5K

GeoDecisions, A Division of Gannett Fleming, Inc.
Camp Hill, Pennsylvania, USA
By Hany Hassaballa, GISP, P.E.

The Center for Geographic Information Systems of the Ministry of Municipality and Environment (CGIS-MME) has produced various topographic maps for more than twenty-five years; however, the agency had never implemented cartographic standards for its map series. In an effort to improve the production and quality of maps for Qatar's citizens, scientists, and officials, CGIS-MME engaged GeoDecisions.

GeoDecisions designed a unique mapping brand and symbol fonts for the maps of the state of Qatar. The new cartography standards needed to meet international color standards and the requirements of national mapping agencies to ensure adherence to best practices. For consistency and quality, the new standards will be applied in the creation of the country's official printed and online maps.

CONTACT
Hany Hassaballa
hhassaballa@geodecisions.com

SOFTWARE
ArcGIS Desktop

DATA SOURCES
The Center for Geographic Information Systems

Courtesy of GeoDecisions.

الدوحة
Doha

SWITZERLAND'S NEW 1:50,000 HIKING MAP

Federal Office of Topography (swisstopo)
Wabern, Switzerland
By Federal Office of Topography swisstopo,
Cartography Division

The updating of the national maps offers new possibilities for the representation of data and is an opportunity to redevelop the well-loved swisstopo hiking maps from scratch. The result is a more homogeneous and readable map, with a modern and colorful look.

The new hiking maps are based on the new generation 1:50,000 scale national map. The data has now been developed in vector form, opening up new possibilities for data use and representation. The thematic content of the hiking maps is therefore no longer just added to the basic map, but instead is integrated with the basic topographical information to create a uniform map. Thanks to the new graphics, the maps are more readable and look fresh and modern.

The trail categories of "hiking trail," "mountain hiking trail," and "alpine hiking trail" are now indicated in yellow, red, and blue, matching the signage on the ground. Labeled public transport stops and pictograms indicating points of interest, observation towers and castles or remote inns, protection huts, parking places, caves and via ferrata provide useful additional information.

CONTACT
Urs Isenegger
urs.isenegger@swisstopo.ch

SOFTWARE
ArcGIS Desktop 10.5

DATA SOURCES
Federal Office of Topography swisstopo

WORLD EXPLORATION

GeoJango
GeoJango.com
Pleasanton, California, USA

GeoJango.com's Voyager 2 World Map was designed to intrigue people to learn more about the world. This map illustrates Earth's beautiful topography as an elaborate piece of art.

The map highlights landmarks including cultural, engineering, and geological sites. The map also includes historical events such as the travels of famous explorers, historical trade routes, and tragic events such as the location of the sinking of the Titanic. Environmental references include the deepest and highest locations on Earth, the hottest recorded location, the most northern city, and the gateway city from South America to Antarctica.

The geography in the map was created by combining multiple satellite imagery resources. The illustration of the terrain was modeled using elevations and vegetation data. A hillshade was applied to illustrate the highest mountains on Earth. The dots in the map can be used to place a push pin to document all one's travels.

CONTACT

Deborah Dennison, GISP
ddennison@geojango.com

SOFTWARE

ArcGIS Desktop, Adobe Illustrator, Adobe Photoshop

DATA SOURCES

GeoJango.com; Shuttle Radar Topography Mission; Moderate Resolution Imaging Spectroradiometer; LandSAT; Natural Earth; National Aeronautics and Space Administration; National Oceanic and Atmospheric Administration; US Geological Survey; United Nations Educational, Scientific and Cultural Organization; CIA World Facts; US National Parks; TripAdvisor; The Nature Conservancy; Wikipedia

Courtesy of GeoJango, LLC.

SALT RIVER WATERSHED MAPPING PROJECT: WILDLIFE HABITAT CORES AND CONNECTIVITY

USACE
Louisville, Kentucky, USA
By Daniel Filiatreau

This map utilizes Esri's Green Infrastructure data to identify important areas of wildlife habitat and networks for wildlife migration within the region of the Salt River Watershed in Kentucky. The map is one in a series of maps completed as part of the Salt River Watershed Conservation Mapping Project, a cooperative effort of over 20 organizations and agencies working to better understand and communicate the public benefits of voluntary conservation measures within the watershed. The participating groups will use the maps to ensure that future projects offer the greatest benefit to the people of the region by supporting clean water, clean air, productive farmland, important wildlife habitat, healthy communities, outdoor recreational opportunities, strong local economies, and the overall desirability of the region as a place to live and work.

CONTACT
Daniel Filiatreau
daniel.l.filiatreau@usace.army.mil

SOFTWARE
ArcGIS Desktop 10.4.1

DATA SOURCES
Esri, US Geological Survey, US Fish and Wildlife Service, US Department of Agriculture, US Census Bureau, Kentucky Department of Transportation, Salt River Watershed Collaborative

Courtesy of Salt River Watershed Collaborative.

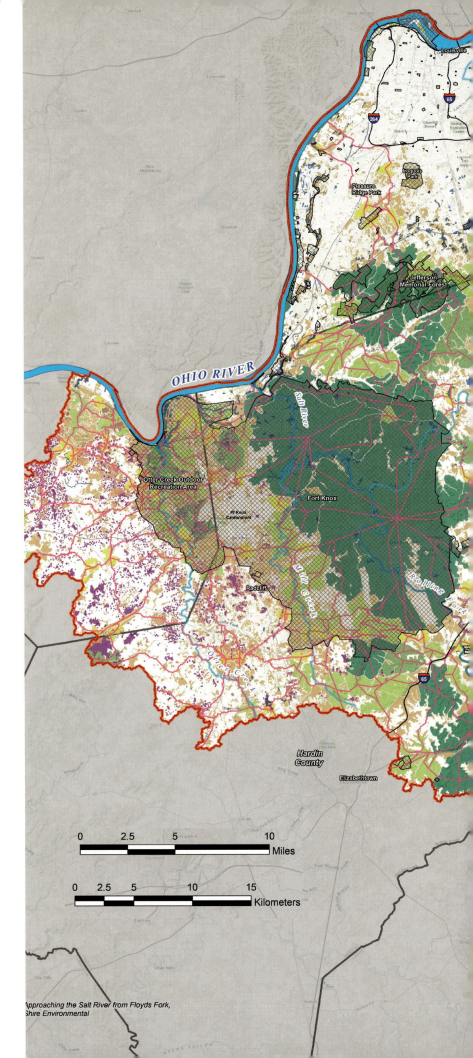

Approaching the Salt River from Floyds Fork, Shire Environmental

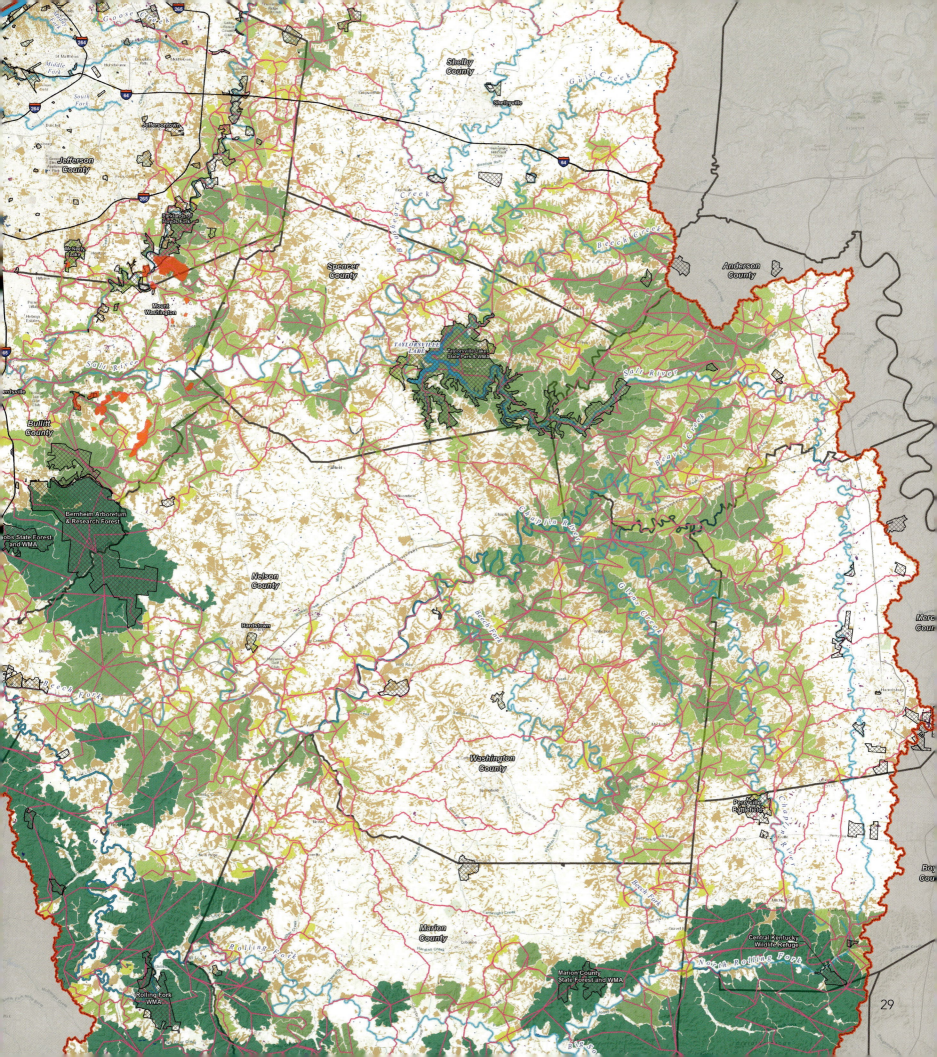

MAINSPRING CONSERVATION TRUST PRIORITIES

Univeristy of Tennessee at Chattanooga
Chattanooga, Tennessee, USA
By Charlie Mix

This map depicts the conservation priorities of Mainspring Conservation Trust, a land trust dedicated to the conservation of the Little Tennessee and Hiwassee River watersheds located in the southern Blue Ridge Mountains of western North Carolina. Mainspring partnered with the University of Tennessee at Chattanooga's Interdisciplinary Geospatial Technology Lab (IGTLab) to develop models that map terrestrial habitat, water resources, recreation and scenic areas, and historic and cultural areas. These models were combined in a weighted overlay creating an index that ranks lands for their conservation value. The final model was used to develop a custom GeoPlanner℠ for ArcGIS application template that Mainspring staff uses to identify focus areas and prioritize projects for conservation and restoration. These data and tools are used daily by Mainspring staff to guide conservation efforts for one of the most biodiverse and culturally rich regions in the US.

CONTACT

Charlie Mix
charles-mix@utc.edu

SOFTWARE

ArcGIS Pro 2.1.2

DATA SOURCES

University of Tennessee Chattanooga IGTLab, US Geological Survey Protected Areas Database, National Conservation Easement Database, Esri, Living Atlas.

Conservation model data sources: North Carolina Natural Heritage Program, North Carolina Division of Water Resources, North Carolina Wildlife Resource Commission, North Carolina State Historic Preservation Office, United States Fish & Wildlife Service National Wetland Inventory, Audubon Society, The Nature Conservancy, United States Forest Service, Appalachian Trail Conservancy, Conservation Trust for North Carolina, North Carolina Onemap, US Geological Survey, Mainspring Conservation Trust

Copyright of the University of Tennessee at Chattanooga & Mainspring Conservation Trust.

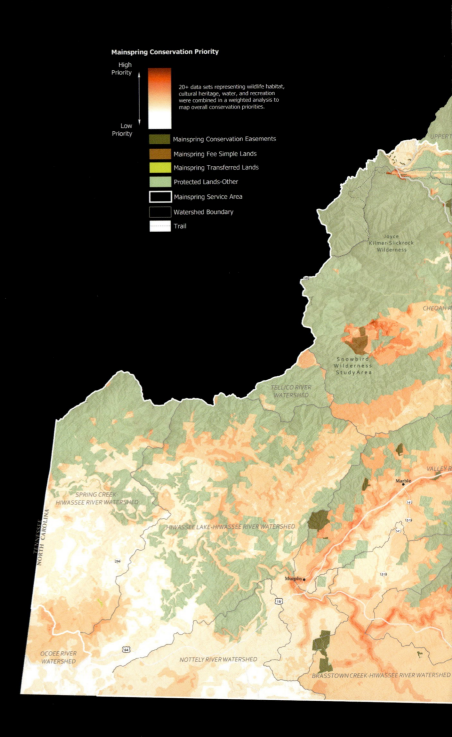

Mainspring Conservation Priority

High Priority

Low Priority

20+ data sets representing wildlife habitat, cultural heritage, water, and recreation were combined in a weighted analysis to map overall conservation priorities.

Mainspring Conservation Easements
Mainspring Fee Simple Lands
Mainspring Transferred Lands
Protected Lands-Other
Mainspring Service Area
Watershed Boundary
Trail

Cartography by Charlie Mix, University of Tennessee Chattanooga-IGTLab, 2018

Projection: NAD83 UTM Zone 17 North

Data Sources: The University of Tennnessee Chattanooga Interdisciplinary Geospatial Technologies Lab, USGS Protected Areas Database, National Conservation Easement Database, Esri Living Atlas

0   5   10   20 Miles

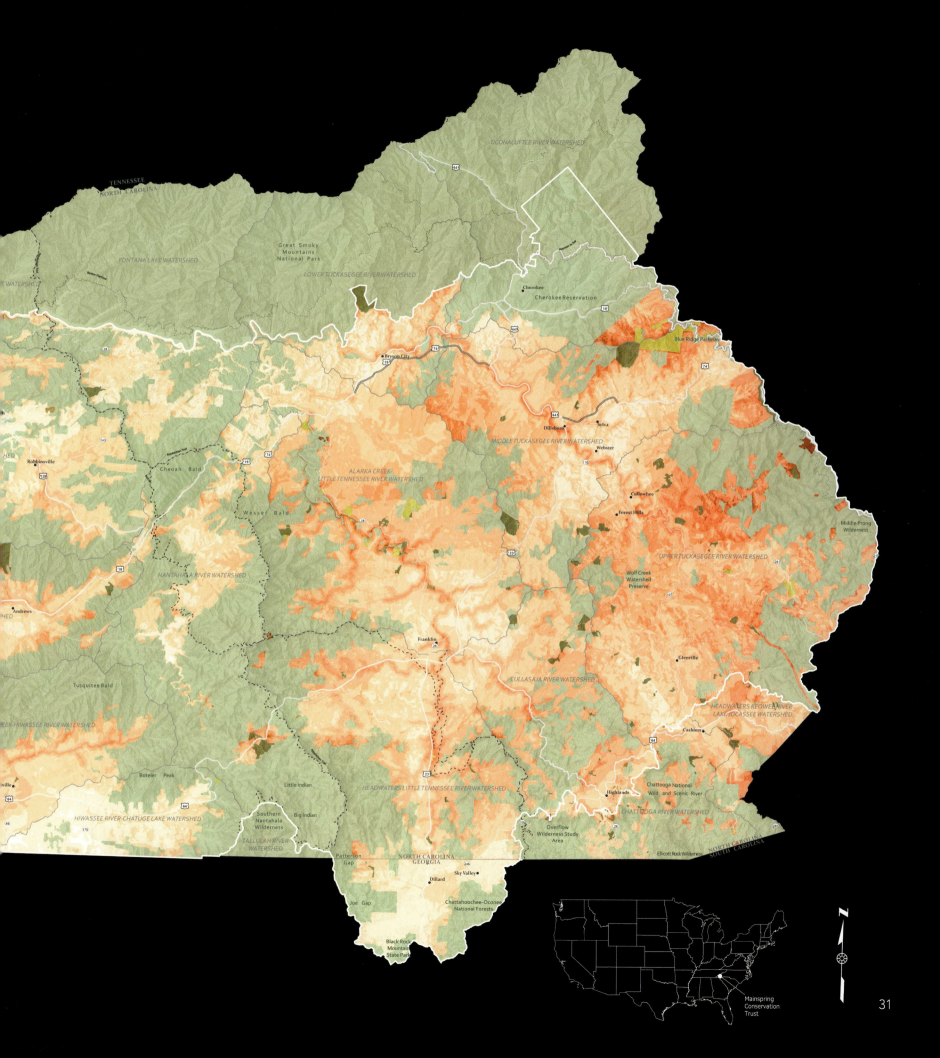

OCONALUFTEE RIVER WATERSHED

TENNESSEE
NORTH CAROLINA

FONTANA LAKE WATERSHED

Great Smoky
Mountains
National Park

LOWER TUCKASEGEE RIVER WATERSHED

Cherokee
Cherokee Reservation

Blue Ridge Parkway

Bryson City

MIDDLE TUCKASEGEE RIVER WATERSHED

Dillsboro   Sylva
Webster

Robbinsville

Cheoah Bald

ALARKA CREEK-
LITTLE TENNESSEE RIVER WATERSHED

Cullowhee

Forest Hills

Wesser Bald

Middle Prong
Wilderness

Andrews

NANTAHALA RIVER WATERSHED

UPPER TUCKASEGEE RIVER WATERSHED

Wolf Creek
Watershed
Preserve

Tusquitee Bald

Franklin

Glenville

CULLASAJA RIVER WATERSHED

HEADWATERS KEOWEE RIVER-
LAKE JOCASSEE WATERSHED

CREEK-HIWASSEE RIVER WATERSHED

Cashiers

Boteler Peak

Little Indian

HEADWATERS LITTLE TENNESSEE RIVER WATERSHED

Highlands

Chattooga National
Wild and Scenic River

HIWASSEE RIVER-CHATUGE LAKE WATERSHED

Southern
Nantahala
Wilderness

Big Indian

CHATTOOGA RIVER WATERSHED

TALLULAH RIVER
WATERSHED

Overflow
Wilderness Study
Area

Ellicott Rock Wilderness

NORTH CAROLINA
SOUTH CAROLINA

Patterson
Gap

NORTH CAROLINA
GEORGIA

Sky Valley

Dillard

Joe Gap

Chattahoochee-Oconee
National Forests

Black Rock
Mountain
State Park

Mainspring
Conservation
Trust

31

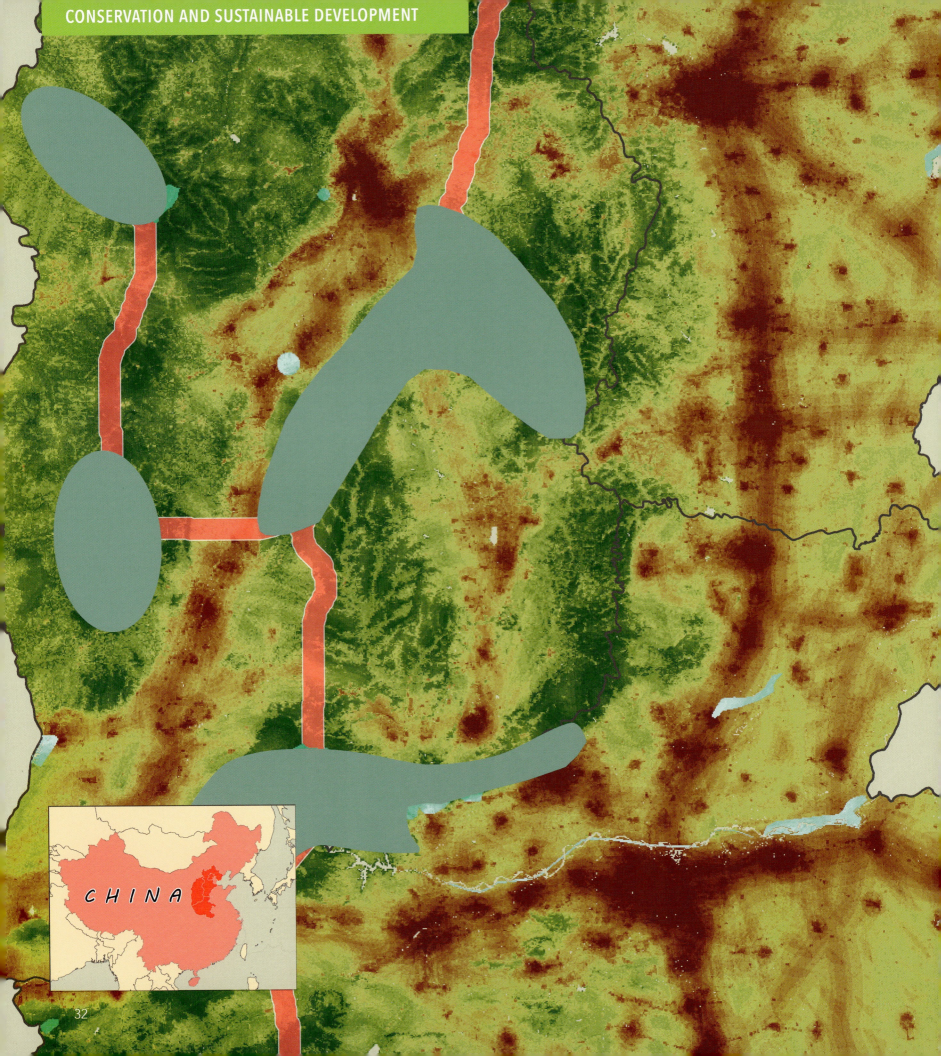

CHINA

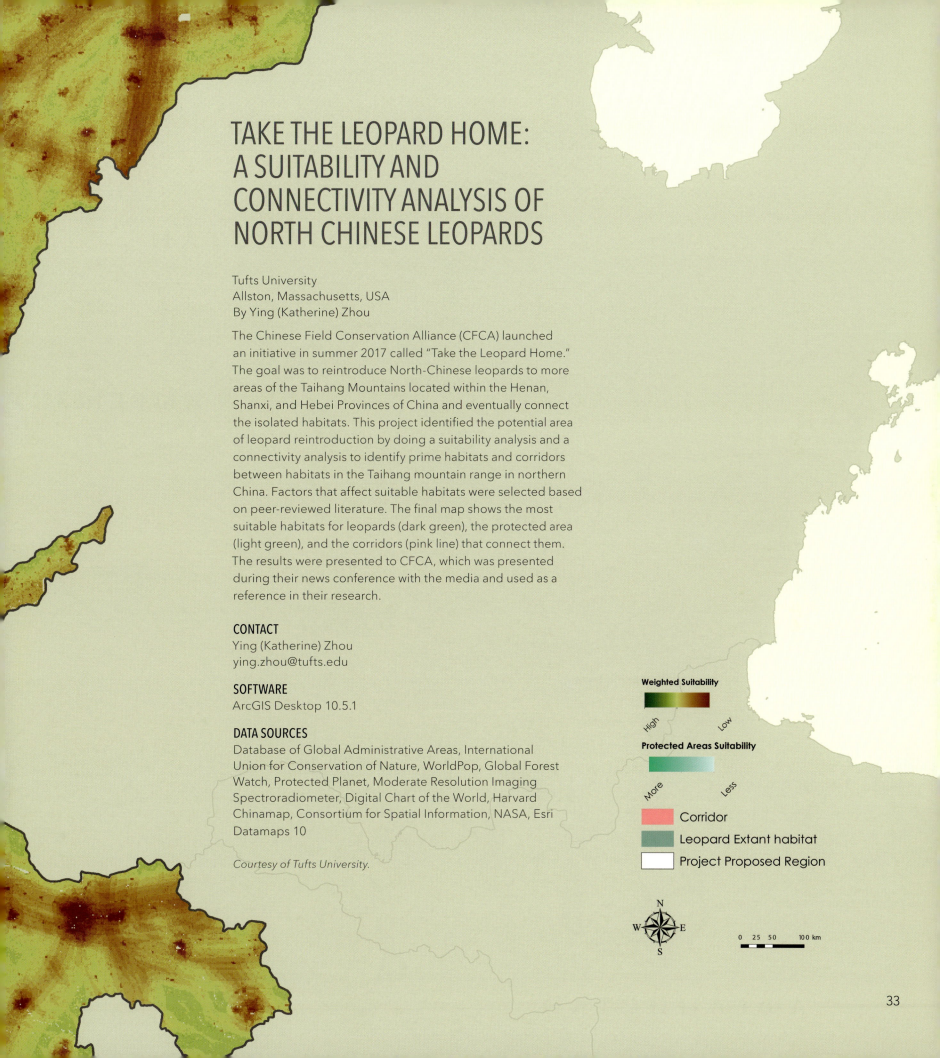

TAKE THE LEOPARD HOME: A SUITABILITY AND CONNECTIVITY ANALYSIS OF NORTH CHINESE LEOPARDS

Tufts University
Allston, Massachusetts, USA
By Ying (Katherine) Zhou

The Chinese Field Conservation Alliance (CFCA) launched an initiative in summer 2017 called "Take the Leopard Home." The goal was to reintroduce North-Chinese leopards to more areas of the Taihang Mountains located within the Henan, Shanxi, and Hebei Provinces of China and eventually connect the isolated habitats. This project identified the potential area of leopard reintroduction by doing a suitability analysis and a connectivity analysis to identify prime habitats and corridors between habitats in the Taihang mountain range in northern China. Factors that affect suitable habitats were selected based on peer-reviewed literature. The final map shows the most suitable habitats for leopards (dark green), the protected area (light green), and the corridors (pink line) that connect them. The results were presented to CFCA, which was presented during their news conference with the media and used as a reference in their research.

CONTACT
Ying (Katherine) Zhou
ying.zhou@tufts.edu

SOFTWARE
ArcGIS Desktop 10.5.1

DATA SOURCES
Database of Global Administrative Areas, International Union for Conservation of Nature, WorldPop, Global Forest Watch, Protected Planet, Moderate Resolution Imaging Spectroradiometer, Digital Chart of the World, Harvard Chinamap, Consortium for Spatial Information, NASA, Esri Datamaps 10

Courtesy of Tufts University.

Weighted Suitability

High   Low

Protected Areas Suitability

More   Less

Corridor
Leopard Extant habitat
Project Proposed Region

0   25   50   100 km

КАZ01 КAZ02

КAZ03

КАЗАХСТАН

КАЗАХСКИЙ
КАРАТАУ

КAZ04
КAZ10
КAZ09   ВЕРХОВЬЯ
                РЕКИ ТАЛАС   Бишкек
Шымкент   КAZ06   КGZ09   КAZ11   КЫРГЫЗСКИЙ
            КAZ05 КAZ   ЗАПАДНЫЙ   КGZ08   АЛАТО (ХРЕБЕ
                   08   ТЯНЬ-ШАНЬ КGZ07
         Регион Ферганской долины   **КЫРГЫЗСТ**
         Смотрите карту-
         вставку   КGZ26   КGZ27

Ташкент   ФЕРГАНСКИЙ
                                                 ХРЕБЕТ

устыня Кызылкум

Айдаркуль UZB21
UZB26   UZB23   UZB22
         UZB24
ОЗЕРО АЙДАРКУЛЬ   КGZ24
И ГОРЫ НУРАТАУ UZB25   КGZ

                                                                              КGZ23

UZB27   UZB
Самарканд   UZB20
         ТAJ04   ТУРКЕСТАНСКИЕ И АЛАЙСКИЕ ГОРЫ
UZB28   ВЕРХОВЬЯ РЕКИ ЗЕРАВШАН
                                      ТAJ05   Смотрите карту-вставку
Карши   UZB29   КGZ21   КGZ22
         ТAJ09 ТAJ06 ТAJ07   Пик Ленина
     Хазрат-Султан   ЦЕНТРАЛЬНЫЙ
     ГОРЫ КОЙТЕНДАГ И ГИССАР ТAJ11   ТAJ08 ТАДЖИКИСТАН   Пик И. Сомони   оз. Каракуль
     UZB31   ТAJ10   ТAJ25   ГОРЫ
         ТAJ12 Душанбе   ТAJ26   +   ПАМИРО-АЛАЯ
ТКМ02   Кафирниган ТAJ21   Пик И. Сомони   И ВАХАН
     UZB32 UZB33   ТAJ24   **ТАДЖИКИСТАН**   ТAJ34   Пик
         UZB36 ТAJ15 ТAJ22   ТAJ27 ТAJ35
ТКМ01 UZB34 ТAJ14 ГОРЫ БАБАТАГ ТAJ23   ПАМИР
М03   UZB35   И КАРАТАУ   ТAJ19   ТAJ28
     ВOДНО-БОЛОТНЫЕ УГОДЬЯ   ТAJ18 ТAJ20   ТAJ32
     ЕЛИФ-ТАЛИМАРДЖАН-ТЕРМЕЗ   ТAJ29 ТAJ30   ТAJ33
                        ТAJ13
                   ТAJ17   AFG01
                        ВЕРХНЯЯ
         ТAJ16   АМУДАРЬЯ
Мазари-Шариф   И ПЯНДЖ   AFG02   ТAJ31

34

АФГАНИСТАН

CONSERVATION OUTCOMES: MOUNTAINS OF CENTRAL ASIA

Conservation International
Arlington, Virginia, USA
By Kellee Koenig

The Critical Ecosystem Partnership Fund's (CEPF) Conservation Outcomes map depicts geographic targets for conservation action within the mountains of central Asia biodiversity hotspot, at the site (Key Biodiversity Area, or KBA) and landscape (conservation corridor) scales.

These targets were defined through a consultative process led by Zoï Environment Network. The hotspot covers 860,000 square kilometers across seven countries and is dominated by the Pamir and Tian Shan mountain ranges. Approximately 64 million people live in the region. Many of the ecosystems long ago reached equilibrium with human activity dominating the landscapes. The region has an established network of protected areas and a tradition of conservation built around respect for natural resources and cultural identification with iconic species.

The map will be in active use until at least 2024. CEPF will use the map to guide the award of grants to civil society organizations working to protect and manage biodiversity in the priority KBAs and corridors throughout the hotspot.

CONTACT
Kellee Koenig
kkoenig@conservation.org

SOFTWARE
ArcGIS Desktop 10.5

DATA SOURCES
Conservation International, Zoï Environment Network, (derivative of) Natural Earth, (derivative of) VMap0, National Geospatial-Intelligence Agency, (derivative of) National Geophysical Data Center, National Oceanic and Atmospheric Administration

Courtesy of Conservation International.

HOW NATURE CAN HELP UGANDA ACHIEVE POLICY GOALS

Conservation International
Arlington, Virginia, USA
By Jenny Hewson, Mariano Gonzalez-Roglich, Kellee Koenig, and David Hunt

Natural ecosystems offer significant potential for mitigation and adaptation to climate change in the form of natural climate solutions (NCS) such as carbon sequestration and storage, provision of freshwater, and protection from storms. Therefore, the value of incorporating NCS into national policy and sustainable development needs to be assessed. In this project, we undertook a pilot study to identify areas of natural capital in Uganda that could help support its policy goals through NCS and to highlight potential actions to ensure the protection of these key areas.

We focused on three sectors identified in Uganda's Nationally Determined Contributions (NDCs) to the Paris Agreement: forestry, wetlands, and agriculture. Using readily-available data and information, we assessed the potential natural capital opportunities by creating replicable workflows to identify priority locations for forest protection and reforestation, wetland restoration, and demarcation, and to help inform areas for which smart agricultural practices may be particularly important. The methodology displayed shows the workflow outlining the process of identifying priority areas for forest protection. The "Additional Results" section shows the priority regions obtained from the analysis in each subsequent sector.

The outputs of this pilot, when combined with local knowledge, could help target policies and processes that will inform a country's NDCs and identify additional NCS opportunities. The same methodology and workflow could be implemented in multiple countries and regions with similar needs to meet their own NDCs or help preserve their ecosystems for future generations.

CONTACT
Kellee Koenig
kkoenig@conservation.org

SOFTWARE
ArcGIS Desktop 10.6.1

DATA SOURCES
Aboveground biomass: updated Baccini et al., 2015 dataset with 30m² spatial resolution, accessibility: travel time to cities in 2015 from Weiss et al., with 1 km² spatial resolution, boundaries, Level 0 Administrative: VMap0 from NGA, modified by CI, deforestation: annual deforestation data from Hansen et al., 2013, with a 30m² spatial resolution, forest cover: 2000 forest cover data from Hansen et al., 2013, with a 30m² spatial resolution, land productivity indicator 2001–2015: Trends.earth, 2017, landcover: Uganda Landcover 2014 Scheme II from Regional Center for Mapping of Resources for Development (RCMRD), with a 30m² spatial resolution, protected Areas: World Database of Protected Areas from UNEP-WCMC, 2017, soil organic carbon: 30cm depth in tons/ha from Hengel et al. (2017) with 250m² spatial resolution

Courtesy of Conservation International.

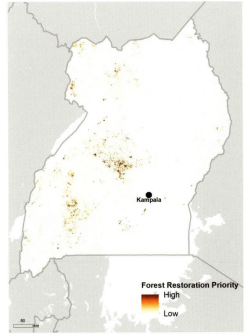

AGB in unprotected forest
(high, medium, low)

Forest Restoration Priority Areas

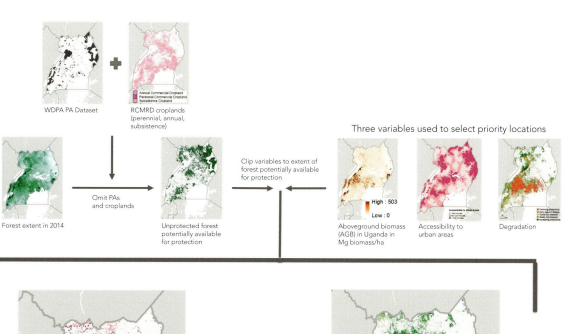

WDPA PA Dataset

RCMRD croplands
(perennial, annual,
subsistence)

Annual Commercial Cropland
Perennial Commercial Cropland
Subsistence Cropland

Forest extent in 2014

Omit PAs
and croplands

Unprotected forest
potentially available
for protection

Clip variables to extent of
forest potentially available
for protection

Three variables used to select priority locations

High : 503

Low : 0

Aboveground biomass
(AGB) in Uganda in
Mg biomass/ha

Accessibility to
urban areas

Degradation

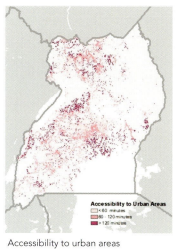

Accessibility to Urban Areas
< 60 minutes
60 - 120 minutes
> 120 minutes

Accessibility to urban areas

+

50 km

high priority for protection
medium priority
low priority

Degradation

=

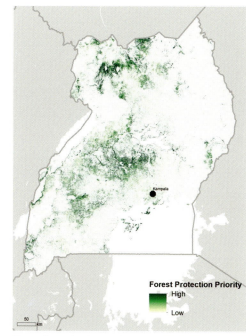

Kampala

Forest Protection Priority
High
Low

50 km

Forest Protection Priority Areas

Additional Results

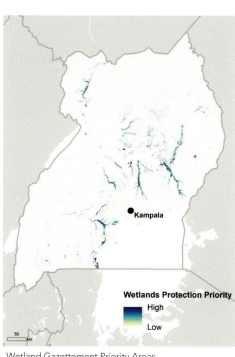

Kampala

Wetland Restoration Priority
High
Low

50 km

Wetland Restoration Priority Areas

Kampala

Wetlands Protection Priority
High
Low

50 km

Wetland Gazettement Priority Areas

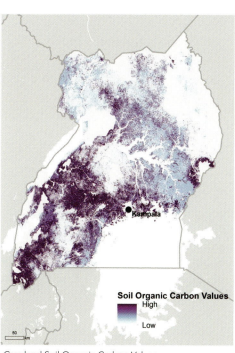

Kampala

Soil Organic Carbon Values
High
Low

50 km

Cropland Soil Organic Carbon Values

STATISTICAL ANALYSIS FOR FINDING OPTIMAL LOCATION OF NEW KINDERGARTENS

MonMap LLC, Ulaanbaatar, Mongolia
By Nyamjargal Sharav

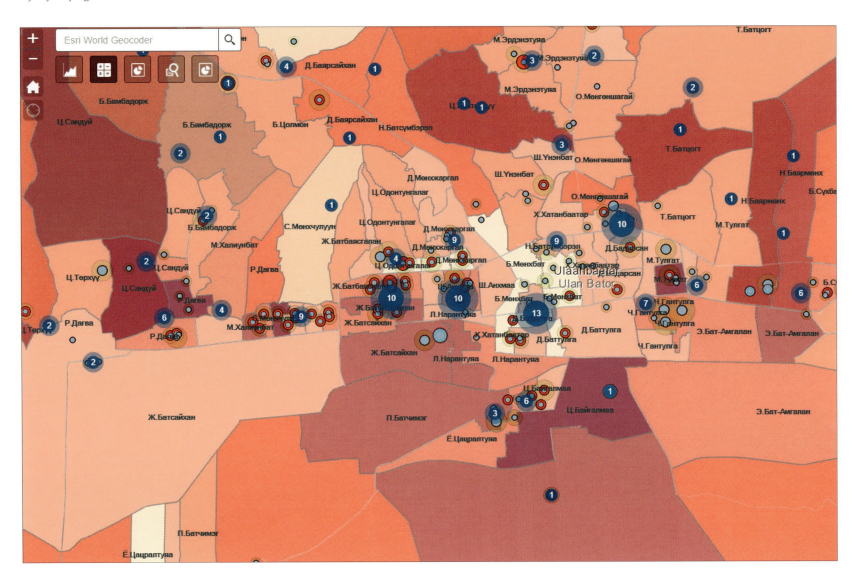

Between 2016 and 2017, various information and data were used to perform the spatial analysis for finding optimal locations for new kindergartens. Points included existing public kindergartens' location and capacity, the current amount of children in each kindergarten, preschool age population by khoroo (micro district), the amount of kindergartens by khoroo, the total number of registered children to enroll in the online application system, the maximum number of new children (two years old and under) in each kindergarten, the locations of newly constructed kindergartens, and representative members' interest area (responsible boundary).

Web applications with spatial and statistical analytics were used to help answer questions about potential new kindergartens and then used to help make optimal decisions for government agencies. Some questions included: Where are the over-capacity kindergartens? Which of micro districts need to build new kindergartens first? How optimal is the location of the proposed new kindergarten?

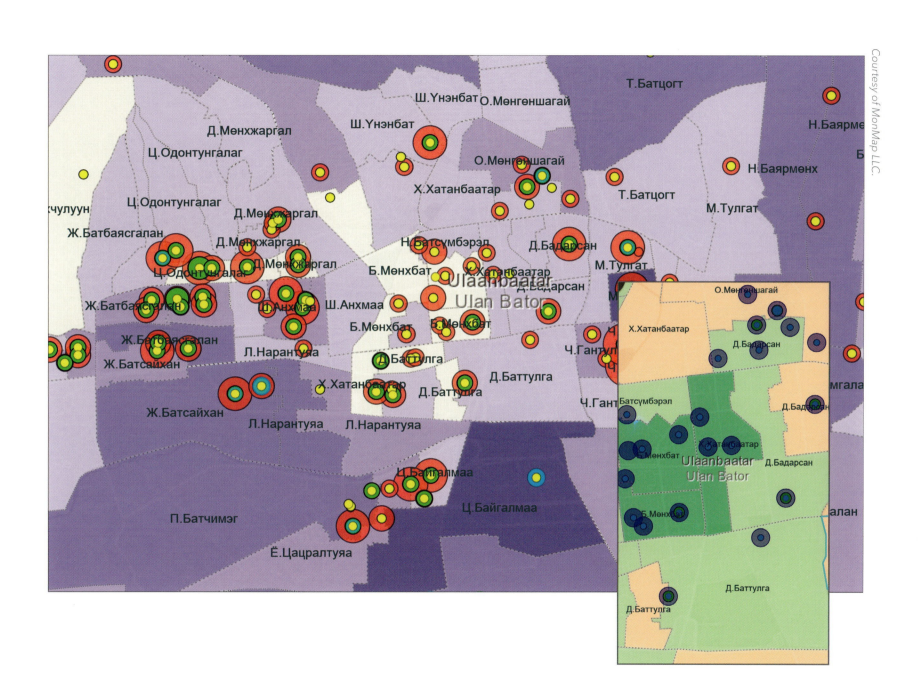

CONTACT
Nyamjargal Sharav
nyamjargal@monmap.mn

SOFTWARE
ArcGIS Online, ArcGIS
Desktop 10.5

DATA SOURCES
Ulaanbaatar Metropolitan Education Department, Master
Planning Agency of Capital City, Communications and
Information Technology Authority, Capital City Statistical Office

ASSESSMENT OF 2017 RECORD FLOOD OF THE SUMMER (AMAN) RICE IN BANGLADESH

FAS/USDA
Washington, District of Columbia, USA
By Arnella Trent and Lisa Colson

Torrential rains on August 11 and 12, 2017, caused landslides and floods throughout Bangladesh, more severe than in the ones experienced in 2015. Although this was the monsoon or wet season, Bangladesh received approximately 60 percent more rainfall than the long-term average in 2017. Rainfall was nearly 270 percent higher in the administrative division, Rajshahi, a major rice producer. What was the impact of these heavy rains on rice production?

Rice is grown throughout the year in Bangladesh, but the Aman season (planted in June and harvested in October) accounts for roughly 40 percent of total production. European Space Agency Sentinel radar analysis was used to compare the 2017 and 2015 floods. Excessive rains impacted roughly 50 percent of the Aman rice area in northern Bangladesh. Rice fields were still flooded near the end of the monsoon which decreased their vegetative response. With 30 days left until harvest, Moderate Resolution Imaging Spectroradiometer (MODIS) normalized difference vegetation index (NDVI) analysis indicated crop health was below average in these flooded areas.

CONTACT
Arnella Trent
arnella.trent@fas.usda.gov

SOFTWARE
ArcGIS Desktop 10.5, Sentinel Application Platform

DATA SOURCES
European Commission, Sentinel 1; National Geospatial-Intelligence Agency, Viznav 2010; The Global Agricultural Monitoring Project NDVI derived from MODIS; World Meteorological Organization; Global Administrative boundaries (modified)

Courtesy of FAS/USDA.

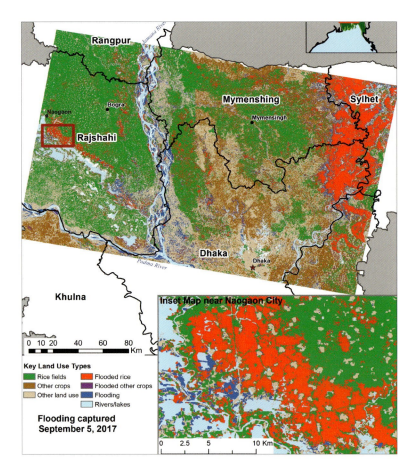

Key Land Use Types

Rice fields	Flooded rice
Other crops	Flooded other crops
Other land use	Flooding
	Rivers/lakes

Flooding captured September 5, 2017

Inset Map near Naogaon City

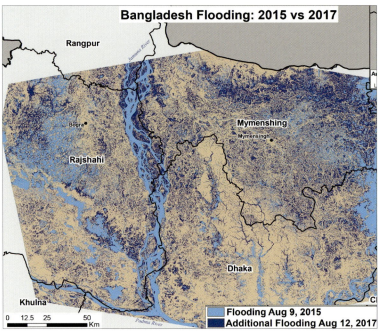

Bangladesh Flooding: 2015 vs 2017

Flooding Aug 9, 2015
Additional Flooding Aug 12, 2017

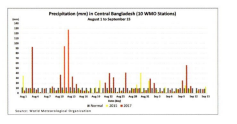

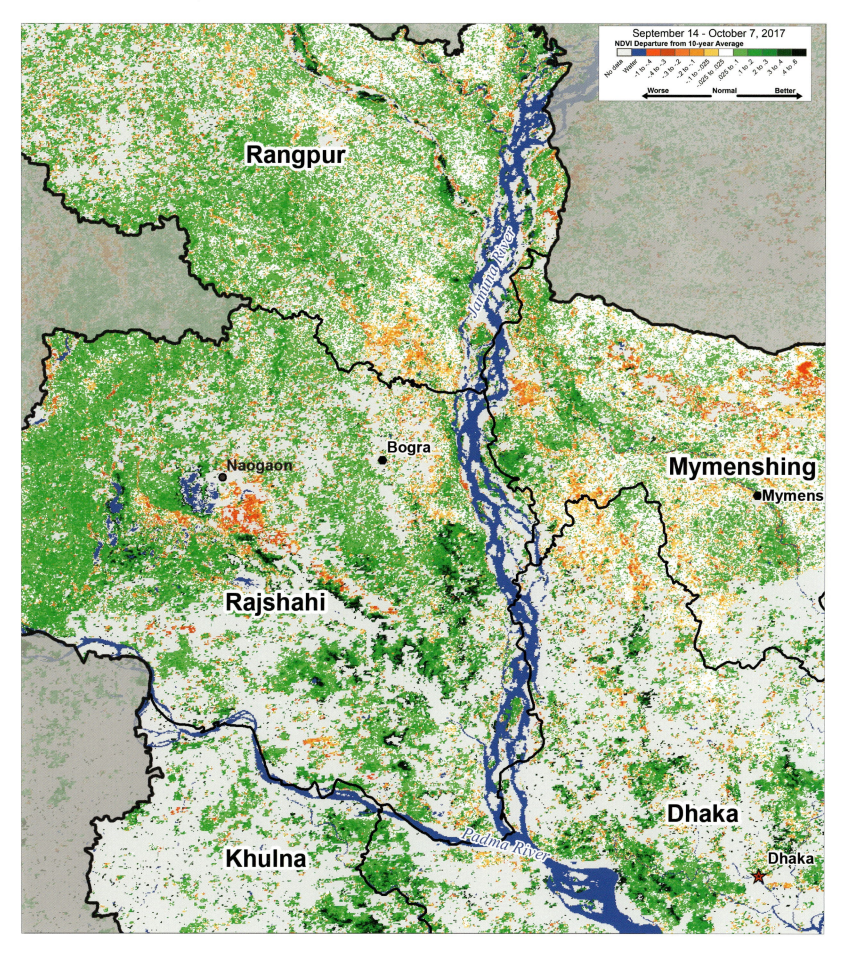

September 14 - October 7, 2017
NDVI Departure from 10-year Average

| No data | Water | -.1 to -.4 | -.4 to -.3 | -.3 to -.2 | -.1 to -.025 | -.025 to .025 | .025 to .1 | .1 to .2 | .2 to .3 | .3 to .4 | .4 to .6 |

Worse ← Normal → Better

Rangpur

Jamuna River

Naogaon

Bogra

Rajshahi

Mymenshing

Mymens

Khulna

Padma River

Dhaka

Dhaka

41

WILDFIRE-INDUCED SOIL EROSION (WISE) MODEL TO PRIORITIZE REMEDIATION EFFORT IN BRITISH COLUMBIA

GeoBC
Victoria, British Columbia, Canada
By Steeve Deschenes

During summer 2017, the province of British Columbia, Canada, experienced its worst recorded wildfire season with more than 1.2 million hectares of forest burned. The large extent of burned forest exposed soil to erosion. Some areas are more at risk than others, and identifying the areas most vulnerable to erosion can improve protection and rehabilitation efforts. The wildfire-induced soil erosion (WISE) model uses burn intensity information derived from satellite images, soil characteristic data, and slope to create an erosion risk index to identify the most critical areas for post-wildfire erosion. The goal of this study was to determine the most sensitive areas for erosion prevention and protection.

The WISE model, as seen on the map, shows the spatial variability of soil erosion risk over the Elephant Hill 2017 wildfire. The model demonstrates the possibility of using spatial information to prioritize remediation and protection efforts to prevent post-wildfire soil erosion. The map shows the location and extent of 106,745 hectares classified as high risk and 7,300 hectares as very high risk, over a total 140,000 hectares burned. The automation of the workflow using Python® is underway to allow post-fire rehabilitation experts to perform the analysis to any area where data is available.

CONTACT
Joy Sinnett
jlsinnett@shaw.ca

SOFTWARE
ArcGIS Desktop 10.3

DATA SOURCES
Building Adaptive & Resilient Communities data, Burned Area Response Team, Province of BC digital elevation model data, Terrain Resource Information Management Soil data, Terrestrial Ecosystem Information, BC Ministry of Environment and Climate Change Strategy

Courtesy of GeoBC.

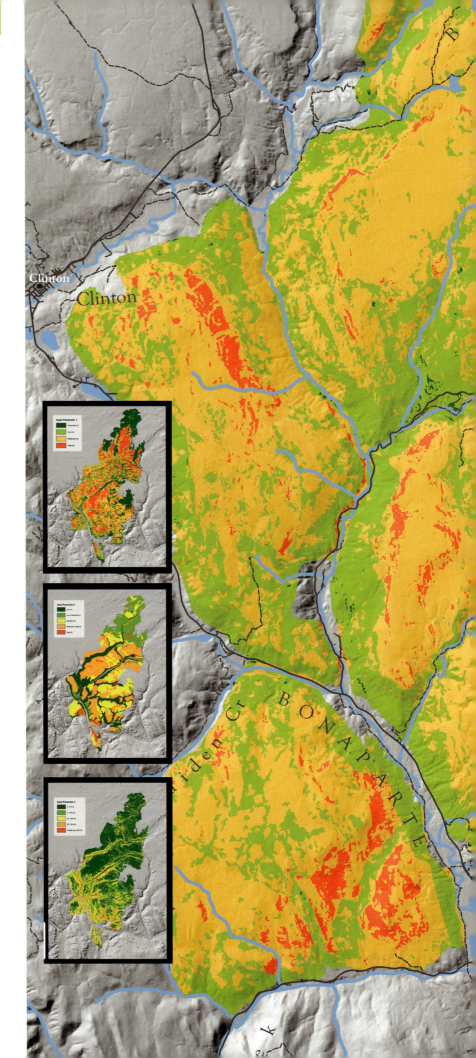

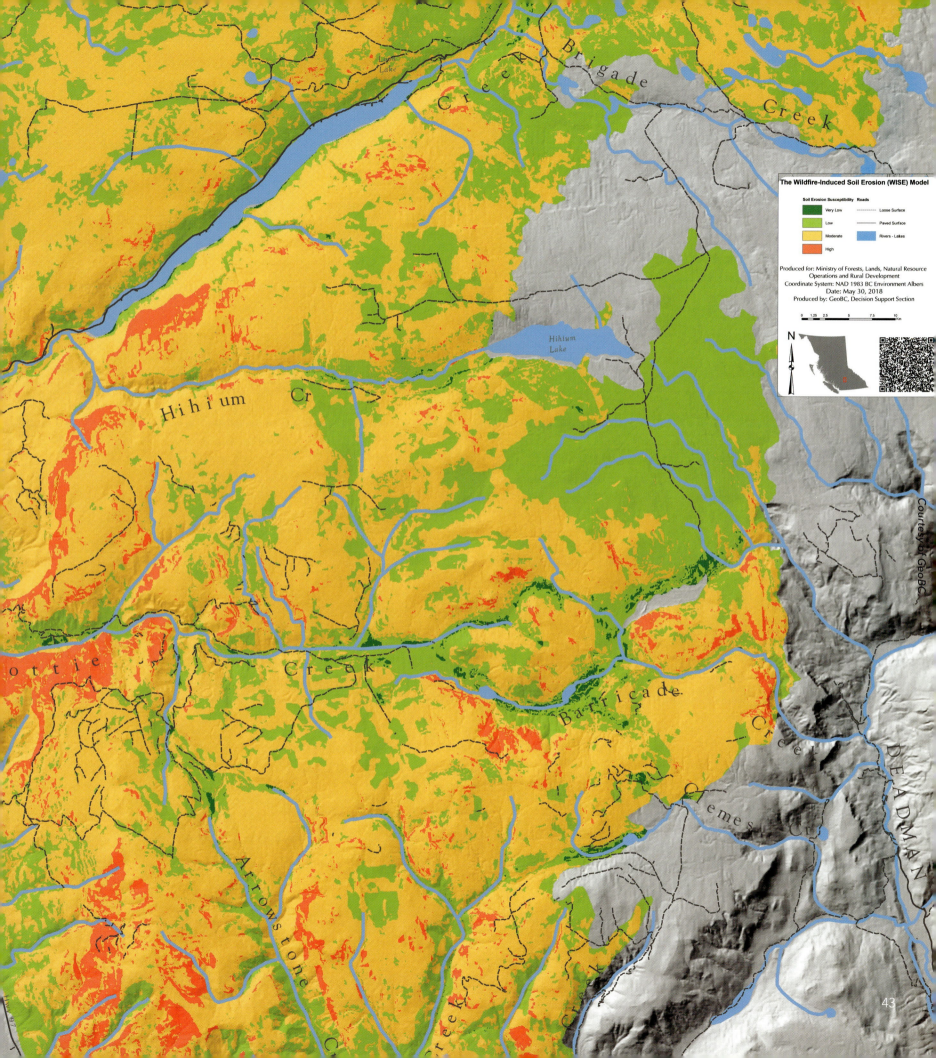

The Wildfire-Induced Soil Erosion (WISE) Model

Soil Erosion Susceptibility
- Very Low
- Low
- Moderate
- High

Roads
- Loose Surface
- Paved Surface

Rivers - Lakes

Produced for: Ministry of Forests, Lands, Natural Resource
Operations and Rural Development
Coordinate System: NAD 1983 BC Environment Albers
Date: May 30, 2018
Produced by: GeoBC, Decision Support Section

0 1.25 2.5   5   10

N

Courtesy of GeoBC

43

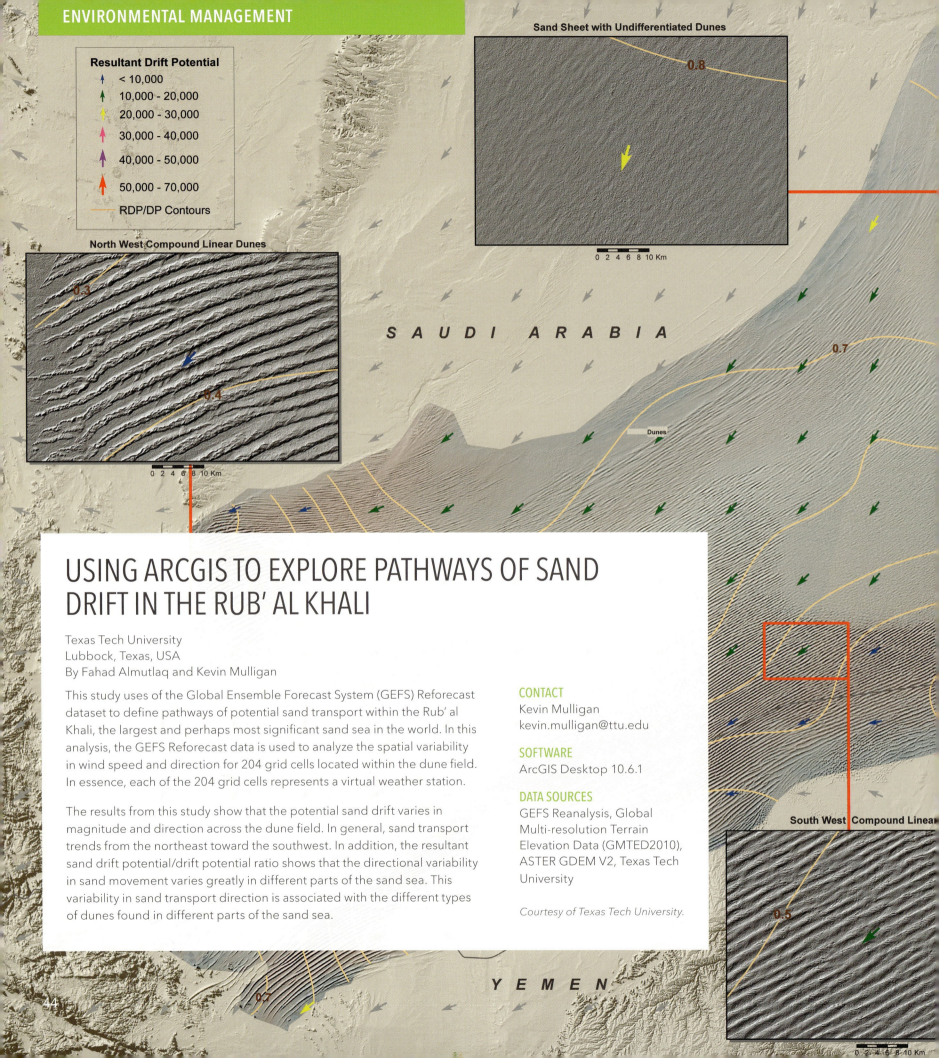

Resultant Drift Potential

- ↑ < 10,000
- ↑ 10,000 - 20,000
- ↑ 20,000 - 30,000
- ↑ 30,000 - 40,000
- ↑ 40,000 - 50,000
- ↑ 50,000 - 70,000
- RDP/DP Contours

Sand Sheet with Undifferentiated Dunes

0.8

0 2 4 6 8 10 Km

North West Compound Linear Dunes

0.3

0.4

0 2 4 6 8 10 Km

S A U D I   A R A B I A

0.7

Dunes

South West Compound Linear

0.5

Y E M E N

0.7

0 2 4 6 8 10 Km

USING ARCGIS TO EXPLORE PATHWAYS OF SAND DRIFT IN THE RUB' AL KHALI

Texas Tech University
Lubbock, Texas, USA
By Fahad Almutlaq and Kevin Mulligan

This study uses of the Global Ensemble Forecast System (GEFS) Reforecast dataset to define pathways of potential sand transport within the Rub' al Khali, the largest and perhaps most significant sand sea in the world. In this analysis, the GEFS Reforecast data is used to analyze the spatial variability in wind speed and direction for 204 grid cells located within the dune field. In essence, each of the 204 grid cells represents a virtual weather station.

The results from this study show that the potential sand drift varies in magnitude and direction across the dune field. In general, sand transport trends from the northeast toward the southwest. In addition, the resultant sand drift potential/drift potential ratio shows that the directional variability in sand movement varies greatly in different parts of the sand sea. This variability in sand transport direction is associated with the different types of dunes found in different parts of the sand sea.

CONTACT
Kevin Mulligan
kevin.mulligan@ttu.edu

SOFTWARE
ArcGIS Desktop 10.6.1

DATA SOURCES
GEFS Reanalysis, Global Multi-resolution Terrain Elevation Data (GMTED2010), ASTER GDEM V2, Texas Tech University

Courtesy of Texas Tech University.

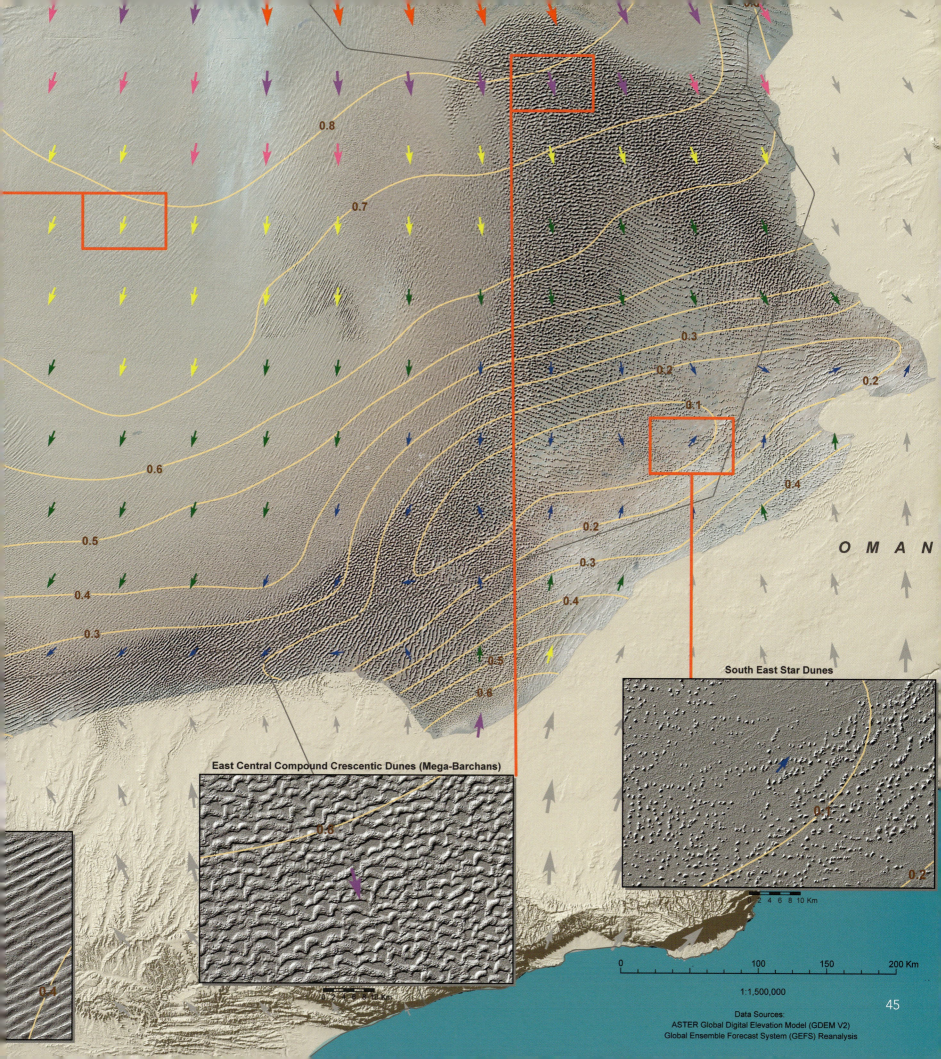

0.8

0.7

0.3
0.2
0.2

0.1

0.2

0.6

0.2

0.5

0.3

0.4

0.4

0.3

O M A N

0.5

0.8

South East Star Dunes

East Central Compound Crescentic Dunes (Mega-Barchans)

0.6

0.6

0.2

0.4

4 6 8 10 Km

0 50 100 150 200 Km

1:1,500,000

45

Data Sources:
ASTER Global Digital Elevation Model (GDEM V2)
Global Ensemble Forecast System (GEFS) Reanalysis

ANTHROPOGENIC BIOMES OF THE WORLD

United Nations Environment Programme
Science Division, Big Data Branch
Nairobi, Kenya
By Jane Muriithi

This map shows the anthropogenic transformation of the terrestrial biosphere from 1700 to 2000.

Somewhere in the range of 1700 and 2000, the earthly biosphere made the critical evolution from generally wild to predominantly anthropogenic, passing the 50 percent mark right off the bat in the twentieth century (Ellis, Klein Goldewijk, Siebert, Lightman & Ramankutty, 2010).

The datasets describe anthropogenic transformations within the terrestrial biosphere caused by sustained direct human interaction with ecosystems, including agriculture and urbanization. Potential natural vegetation, or biomes such as tropical rainforests or grasslands, are based on global vegetation patterns related to climate and geology.

Anthropogenic transformations within each biome is approximated using population density, agricultural intensity (cropland and pasture), and urbanization. This dataset is part of a time series for the years 1700, 1800, 1900, and 2000 that provides global patterns of historical transformation of the terrestrial biosphere during the Industrial Revolution.

CONTACT
Jane Muriithi
jane.muriithi@un.org

SOFTWARE
ArcGIS Desktop 10.6

DATA SOURCES
Socioeconomic Data and Applications Center

Courtesy of SEDAC and United Nations Environment Programme.

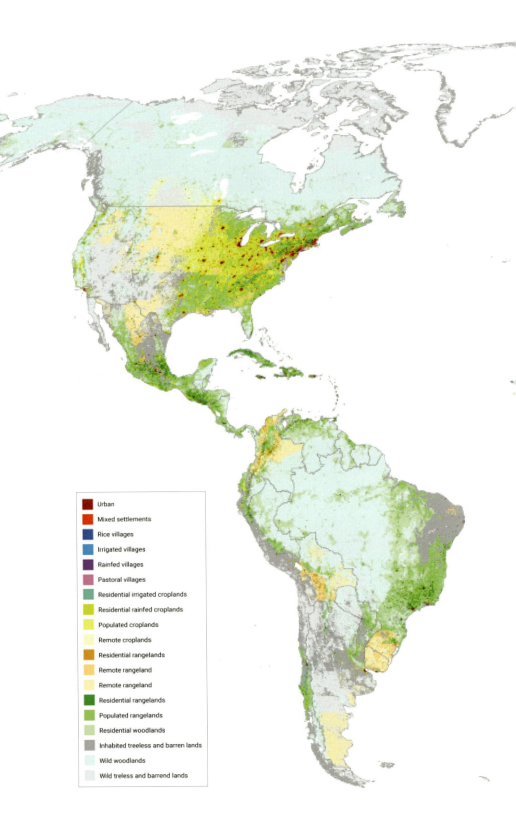

Urban
Mixed settlements
Rice villages
Irrigated villages
Rainfed villages
Pastoral villages
Residential irrigated croplands
Residential rainfed croplands
Populated croplands
Remote croplands
Residential rangelands
Remote rangeland
Remote rangeland
Residential rangelands
Populated rangelands
Residential woodlands
Inhabited treeless and barren lands
Wild woodlands
Wild treless and barrend lands

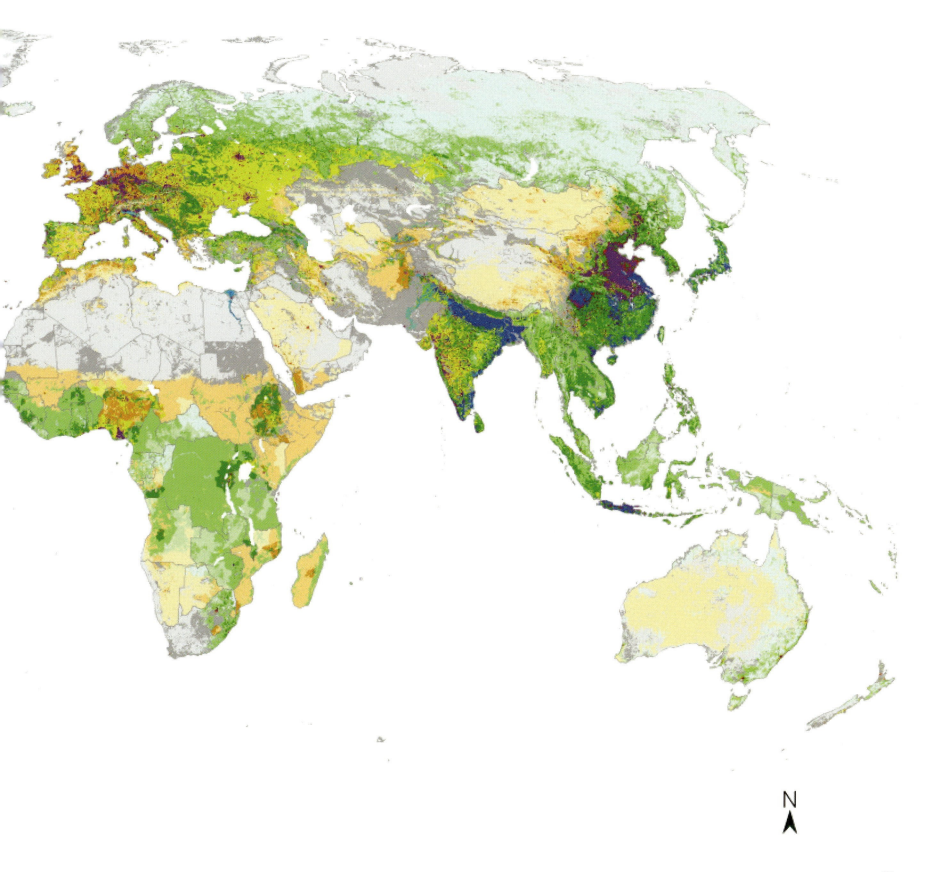

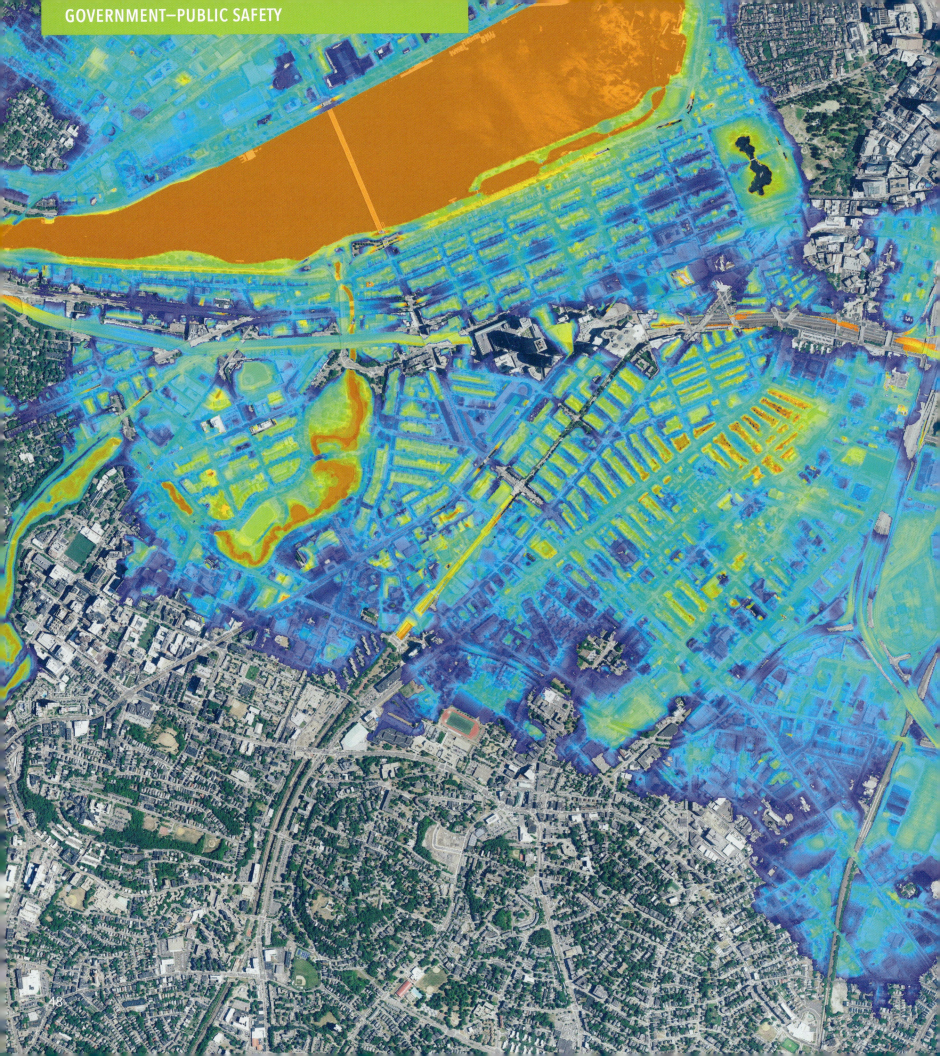

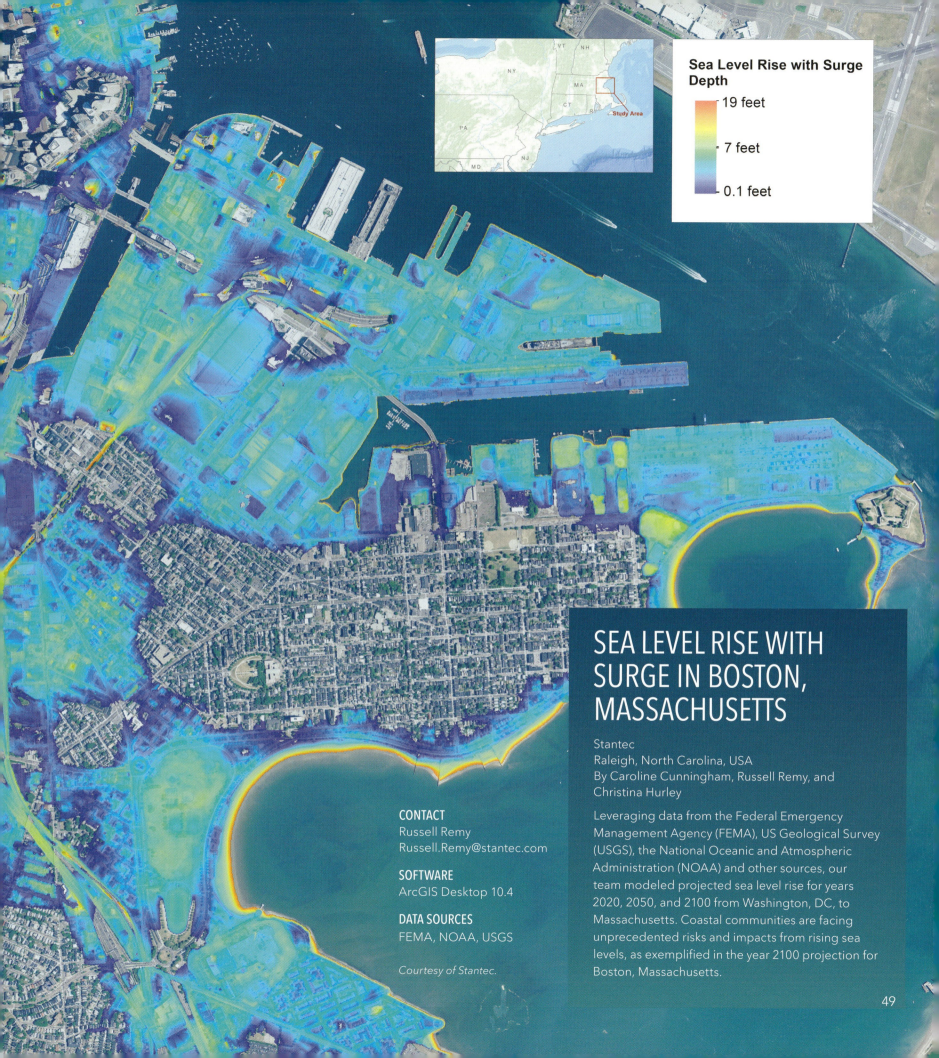

Sea Level Rise with Surge Depth

19 feet

7 feet

0.1 feet

SEA LEVEL RISE WITH SURGE IN BOSTON, MASSACHUSETTS

Stantec
Raleigh, North Carolina, USA
By Caroline Cunningham, Russell Remy, and Christina Hurley

Leveraging data from the Federal Emergency Management Agency (FEMA), US Geological Survey (USGS), the National Oceanic and Atmospheric Administration (NOAA) and other sources, our team modeled projected sea level rise for years 2020, 2050, and 2100 from Washington, DC, to Massachusetts. Coastal communities are facing unprecedented risks and impacts from rising sea levels, as exemplified in the year 2100 projection for Boston, Massachusetts.

CONTACT
Russell Remy
Russell.Remy@stantec.com

SOFTWARE
ArcGIS Desktop 10.4

DATA SOURCES
FEMA, NOAA, USGS

Courtesy of Stantec.

EMERGENCY SPATIAL SUPPORT CENTER–HOW INDONESIA IS RESPONDING TO DISASTER USING SPATIAL TECHNOLOGY

Esri Indonesia
Jakarta, Indonesia
By Ahmad Muttaqin Alim

Emergency Spatial Support Center (ESSC) together with Muhammadiyah Disaster Management Center (MDMC) assist the ministries, agencies, and the public in providing accurate information and data related to disaster in Indonesia in the form of geospatial-based web application solutions. When a disaster occurs, they respond immediately to collaborate information and data from any reliable sources, then quickly develop various thematic applications using the ArcGIS platform.

All of these applications are developed using ArcGIS Online platform with data sourced from relevant ministries and agencies, which has been published for the public. The thematic disaster applications that have been developed are the themes of disasters that cause serious impact in Indonesia.

CONTACT
Khairul Amri
kamri@esriindonesia.co.id

SOFTWARE
ArcGIS Online

DATA SOURCES
MDMC, Indonesian Ministries, Indonesian Agencies

Courtesy of Esri Indonesia.

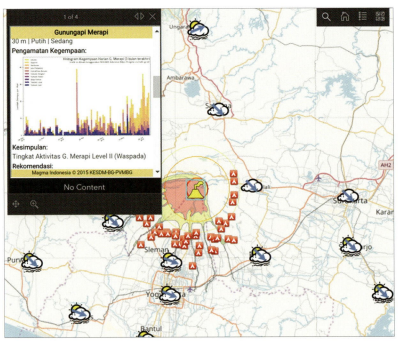

Gn. Merapi Realtime Monitoring

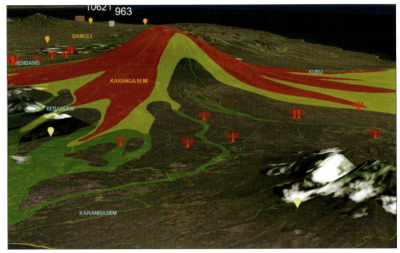

Disaster-Prone Areas (3D)

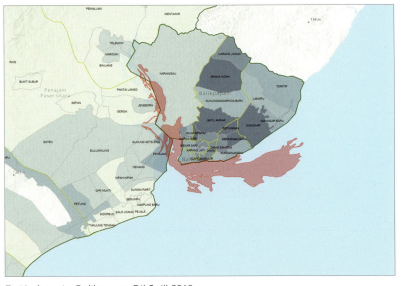

Esri Indonesia–Balikpapan Oil Spill 2018

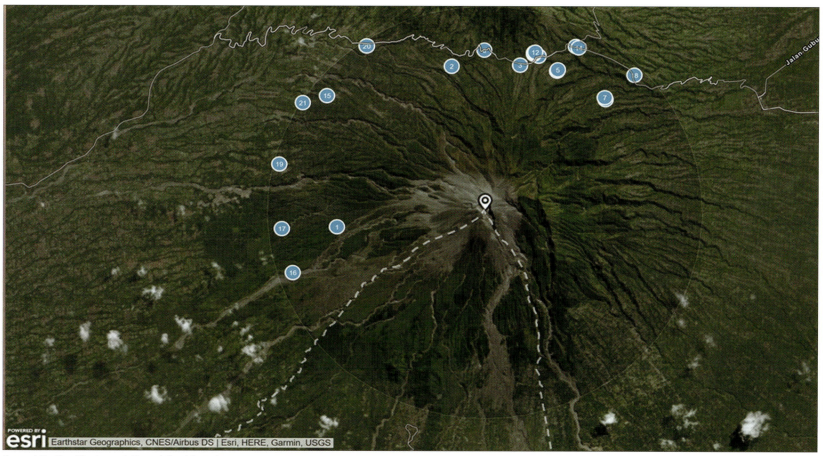

Potensi Faslitas Terdampak Erupsi Gn. Merapi

Esri Indonesia–Balikpapan Oil Spill 2018

Dashboard Survey Gempa Banjarnegara
(Banjarnegara Earthquake Damage Assessment)

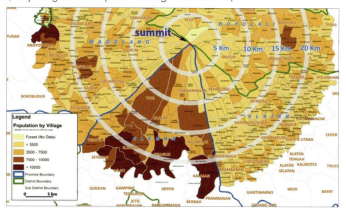

Population Surrounding Mt. Merapi

TARGET HAZARD ANALYSIS IN SIOUX FALLS, SD

City of Sioux Falls
Sioux Falls, South Dakota, USA
By Austin Brynjulson

Sioux Falls Fire Rescue (SFFR) has been an avid consumer of GIS for nearly thirty years. In order to meet the requirements of NFPA (National Fire Protection Association) 1620: Standard for Pre-Incident Planning, SFFR determined a need to utilize Target Hazard Analysis, a solution from Esri that can be used by GIS and fire personnel to aid in identifying vulnerable and high-risk properties and to establish a list of buildings that warrant a pre-incident plan. Pre-incident plans are developed to provide building feature information during emergencies at high hazard occupancies.

To calculate target hazards, the city of Sioux Falls GIS staff utilized building footprint and parcel data to calculate six different criticality scores based on a property's occupancy type, economic impact, fire flow, building height, building area, and life safety category. The higher the criticality score, the greater the risk associated with that property. The scores for the six criteria were combined with the city's address data and used to determine a total hazard score for all addresses in Sioux Falls, as shown in this map. Further analysis can then be done across the city to determine density or significant clusters of high-risk properties.

CONTACT

Austin Brynjulson
ABrynjulson@siouxfalls.org

SOFTWARE

ArcGIS Pro 2.1.2

DATA SOURCES

City of Sioux Falls

Courtesy of City of Sioux Falls.

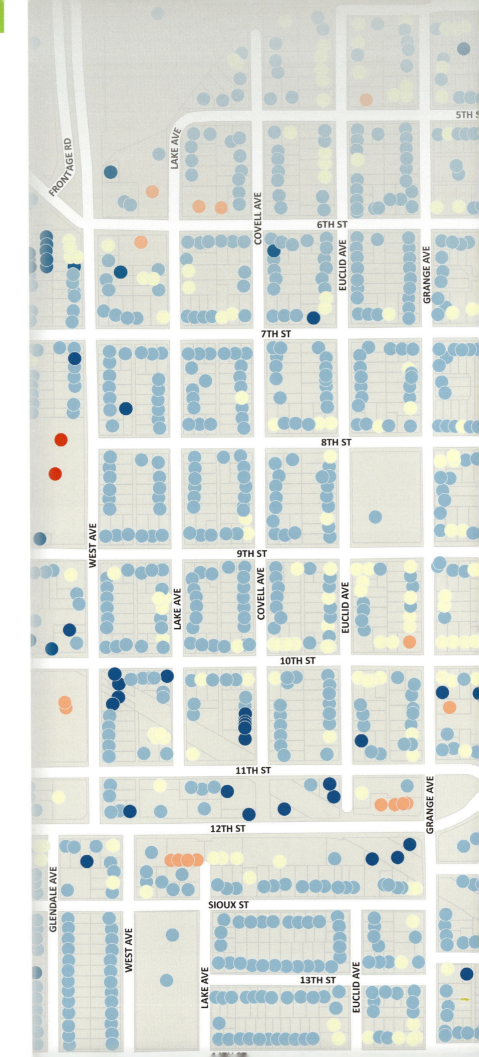

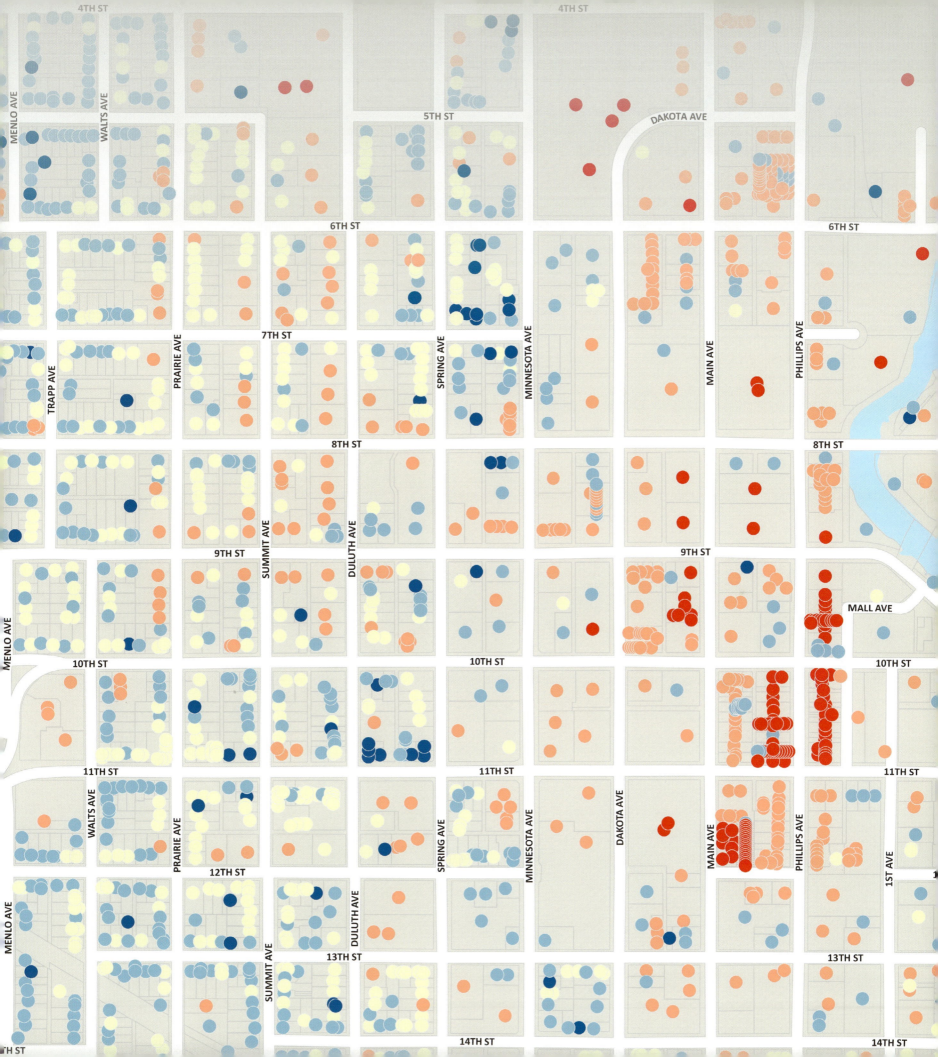

FIRE ANALYSIS OF THE THOMAS FIRE IN CALIFORNIA USING NASA DATA IN A GIS

NASA
Goddard Space Flight Center, Maryland, USA
By Ross Bagwell, Byron Peters, and Minnie Wong

These maps are an analysis of the Thomas Fire that occurred in California during December 2017. Using a variety of NASA Earth science data from five National Aeronautics and Space Administration (NASA) sources (including four Earth Observing System Data and Information System Distributed Active Archive Centers and NASA Fire Information for Resource Management System), as well as ancillary data from Ventura County, Santa Barbara County, and the Department of Homeland Security, this analysis sought to identify forest fire risk zones, create a fire occurrence density map, examine the vegetation and subsequent burn scar, capture the affected parcels, and capture the affected vegetation.

CONTACT
Ross Bagwell
ross.bagwell@nasa.gov

SOFTWARE
ArcGIS Desktop 10.5

DATA SOURCES
NASA

Courtesy of NASA.

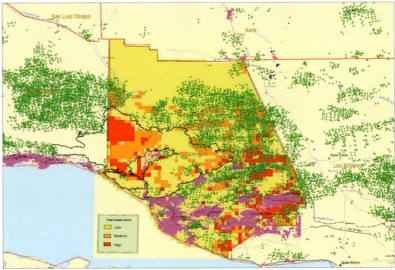

NASA Firms hotspots in Ventura and Santa Barbara counties

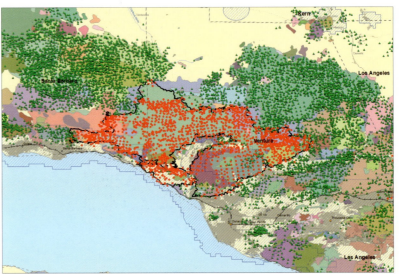

NASA FIRMS Active Fire Hotspots in Ventura and Santa Barbara Counties

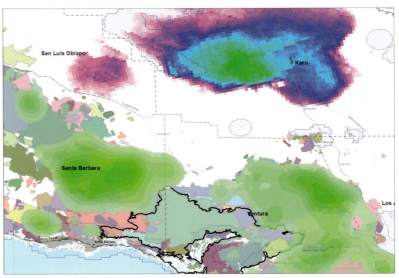

NASA Daymet Monthly Precipitation Data

NASA Normalized Difference Vegetation Index (NDVI)

Copernicus Sentinel-1A Synthetic Aperture Radar (SAR) data, retrieved on January 2018

Enhanced Vegetation Index (EVI)

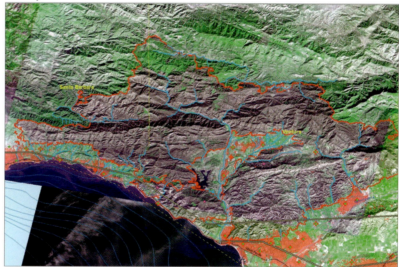

NASA Aster L1T data, captured on December 26, 2017

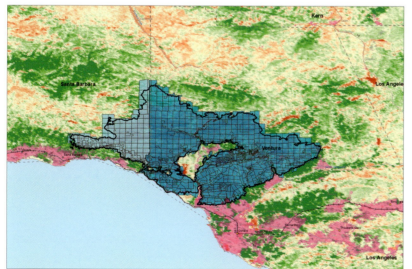

Intersection of the Thomas Fire containment boundary

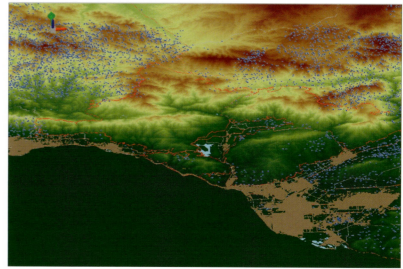

NASA Global Digital Elevation Model (GDEM)

FLOOD ANALYSIS OF THE THOMAS FLOODS USING NASA DATA

NASA/SSAI
Greenbelt, Maryland, USA
By Byron Peters and Ross Bagwell

This analysis is a follow-on to the Thomas Fire analysis presented by Ross Bagwell ("Fire Analysis of the Thomas Fire in California Using NASA Data in a GIS"). The Thomas Fire and heavy rains a month later led to the historic flooding. The maps tell the story using National Aeronautic and Space Administration (NASA) Earth Observing System data in concert with Santa Clara County data.

CONTACT

Byron Peters
Byron.V.Peters@nasa.gov

SOFTWARE

ArcGIS Desktop 10.6

DATA SOURCES

NASA Land Processes Distributed Active Archive Center, NASA Alaska Satellite Facility Distributed Active Archive Center, Santa Clara County, NASA Distributed Active Archive Centers

Courtesy of the NASA ESDIS Project.

Enhanced Vegetation Index (EVI)

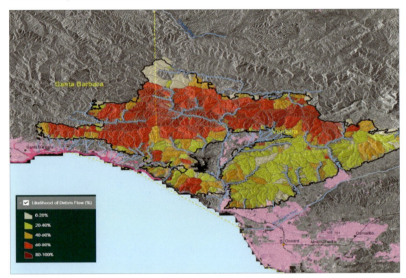

Copernicus Sentinel-1A Synthetic Aperture Radar (SAR) data

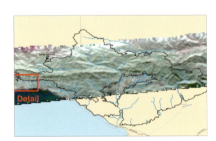

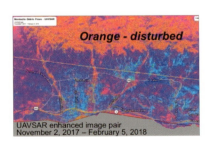

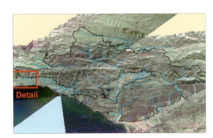

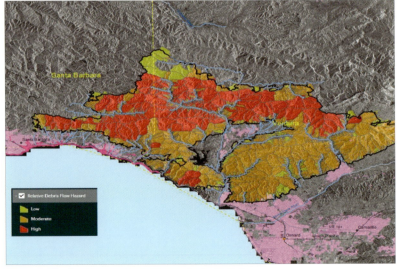

Copernicus Sentinel-1A Synthetic Aperture Radar (SAR) data

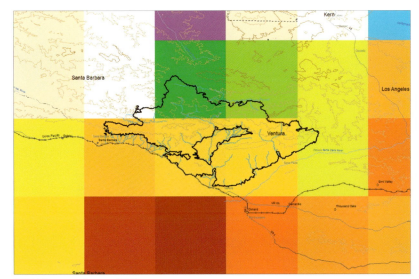

GPM Microwave Imager (GMI)

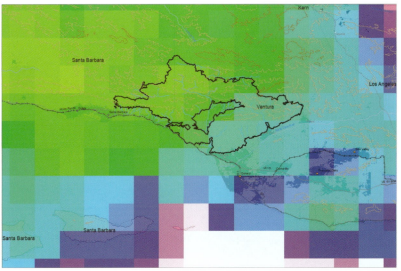

NASA IMERG data

NASA Global Digital Elevation Model (GDEM)

Thomas Fire burn scar outline with major contour lines

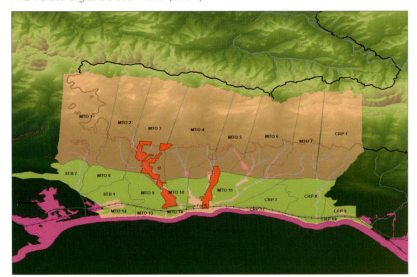

Zooming in closer to the Santa Barbara area within the NASA GDEM

NASA Soil Moisture Active Passive (SMAP) data

Forest Grove
Cornelius
Hillsboro
Beaverton
Tigard
King City
Lake Oswego
Tualatin
West Linn
Sherwood
Oregon
Wilsonville
Portland
Milwaukie
Canby

- ● Commercial property
- ● Industrial property
- ● Multifamily residential
- ● Multifamily with on-site commercial
- · Single family homes
- ☐ Current Urban Growth Boundary (UGB)
- ☐ UGB expansion prior to 2007
- ☐ UGB expansion 2007 and later

Units built over time by housing type

MUR- Multifamily with on-site commercial
MFR- Other multifamily
SFR- Single family

LAND DEVELOPMENT MONITORING SYSTEM FOR THE PORTLAND REGION

Metro Data Resource Center
Portland, Oregon, USA
By Al Mowbray

In 2015, Metro's Data Resource Center (DRC) began formulating a land development monitoring system (LDMS) to examine development trends within the urban growth boundary (UGB) over time, enabled by the twenty-plus years of published GIS data produced by the DRC. The LDMS identifies land parcels developed between 2007 and 2016, and both categorizes and quantifies the changes identified. This map illustrates LDMS findings pertaining to residential, commercial, and industrial development, and application of those findings to improve land use forecasts.

The primary input data layers used for the LDMS are the current and historical tax lot parcels (provided by each county jurisdiction), the vacant land inventory (VLI, from visual interpretation of annually acquired aerial imagery since 1996), and the multifamily housing database (from a variety of sources compiled by Metro annually since 2010). These and many other data layers are part of the Regional Land Information System (RLIS), an ever-expanding catalog of map layers updated quarterly by the DRC since 1996.

CONTACT
Al Mowbray
al.mowbray@oregonmetro.gov

SOFTWARE
ArcGIS Pro 2.2

DATA SOURCES
Oregon Metro RLIS

Courtesy of Metro Data Resource Center.

FUTURE STATE PLANE COORDINATE SYSTEM FOR THE UNITED STATES

NOAA's National Geodetic Survey
Silver Spring, Maryland, USA
By Michael L. Dennis, Dana J. Caccamise II, and William A. Stone

The State Plane Coordinate System (SPCS) of the United States will soon change: National Geodetic Survey (NGS) from the National Oceanic and Atmospheric Administration (NOAA) will create a new version—SPCS2022—to replace the existing North American Datum of 1983 (NAD 83) version, SPCS 83. This change will accompany the transition from NAD 83 to the Terrestrial Reference Frames of 2022 as part of an overall modernization of the US National Spatial Reference System.

Because the change will significantly impact US mapping, surveying, engineering, and myriad georeferenced activities, NGS is seeking user-community input on the development of SPCS2022. In support of this process—and to help state stakeholders make informed decisions on designs that best meet their needs—NGS is creating preliminary designs of SPCS2022 zones, including new statewide zones (subject to change, based in part on state input).

Examples of these SPCS2022 designs are illustrated here, along with maps of existing SPCS 83 zones for comparison. The maps show linear distortion (map scale error) at the topographic surface, along with various performance statistics. This information allows visual and quantitative assessment of the proposed SPCS2022 zones. The intent is for SPCS2022 to be a technically sound and practical component of the nation's future spatial data infrastructure, fully satisfying the broad needs and applications of the geospatial community for years to come.

CONTACT
Michael Dennis
michael.dennis@noaa.gov

SOFTWARE
ArcGIS Desktop 10.5.1

DATA SOURCES
NGS, NOAA, US Geological Survey, Esri

Courtesy of NOAA's National Geodetic Survey.

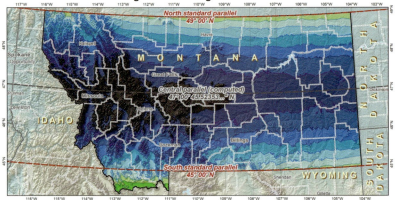

Existing SPCS 83 design: **Montana Zone**

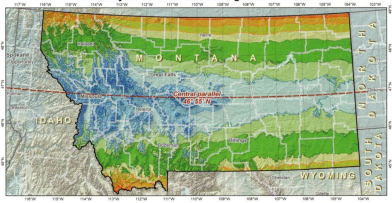

Preliminary SPCS2022 default design: **Montana Zone**

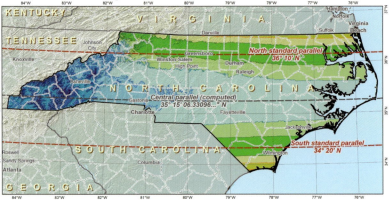

Existing SPCS 83 design: **North Carolina Zone**

Preliminary SPCS2022 default design: **North Carolina Zone**

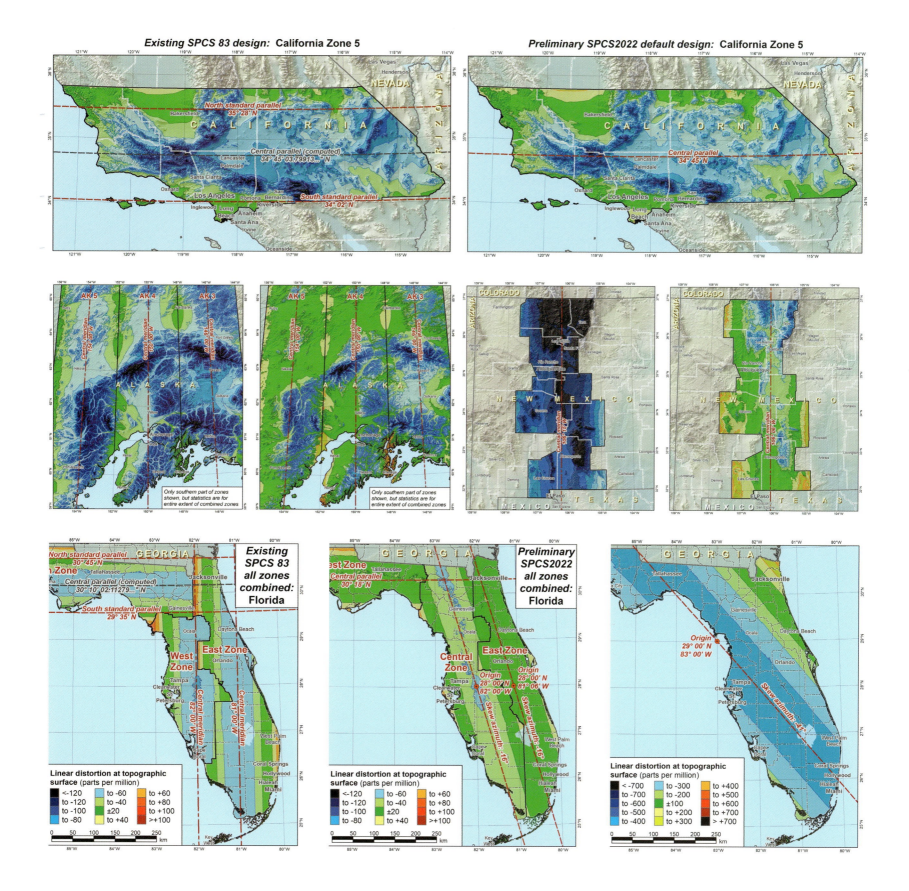

Existing SPCS 83 design: California Zone 5

Preliminary SPCS2022 default design: California Zone 5

Existing SPCS 83 all zones combined: Florida

Preliminary SPCS2022 all zones combined: Florida

Linear distortion at topographic surface (parts per million)

<-120	to -60	to +60
to -120	to -40	to +80
to -100	±20	to +100
to -80	to +40	>+100

Linear distortion at topographic surface (parts per million)

<-120	to -60	to +60
to -120	to -40	to +80
to -100	±20	to +100
to -80	to +40	>+100

Linear distortion at topographic surface (parts per million)

< -700	to -300	to +400
to -700	to -200	to +500
to -600	±100	to +600
to -500	to +200	to +700
to -400	to +300	> +700

AUTOMATED MAP PRODUCTION IN THE CLOUD

Ordnance Survey
Southampton, United Kingdom
By Samuel Andrews, Tom Sutcliffe, and Derek Howland

This map demonstrates the results of a project completed by Ordnance Survey, the United Kingdom's National Mapping Agency, to automate the production of a 1:10,000 scale contextual mapping product, VectorMap Local. VectorMap Local is Ordnance Survey's premium local scale mapping product used by businesses and government agencies for a variety of use cases including emergency response, urban planning, and day-to-day mapping. The aim of the project was to replace the increasingly unreliable, existing manual production system with a fully automated solution.

The map of Durham City Centre is split into two halves. On the left is a map tile created by the older production system where aspects such as feature generalization and label placement are manually handled by a team of cartographers. On the right is a map tile created to the same specification using the new fully-automated cloud-based production system. In the new system, overall orchestration is handled by Esri Workflow Manager and Python scripts. Feature generalization is done by leveraging the Esri toolset, using a series of Esri ModelBuilder models. Map styling and label placement is stored across over 100 symbol and label classes in an MXD file. A fully national set can be processed and published in just 10 days.

CONTACT
Samuel Andrews
Samuel.Andrews@os.uk

SOFTWARE
ArcGIS Desktop 10.6

DATA SOURCES
Ordnance Survey source data

Contains OS data. Crow copyright and database right 2018.

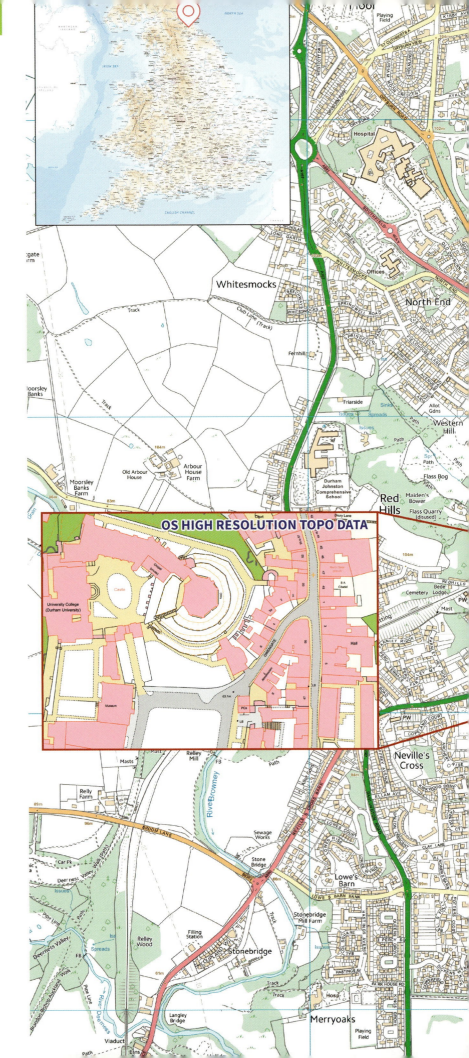

OS HIGH RESOLUTION TOPO DATA

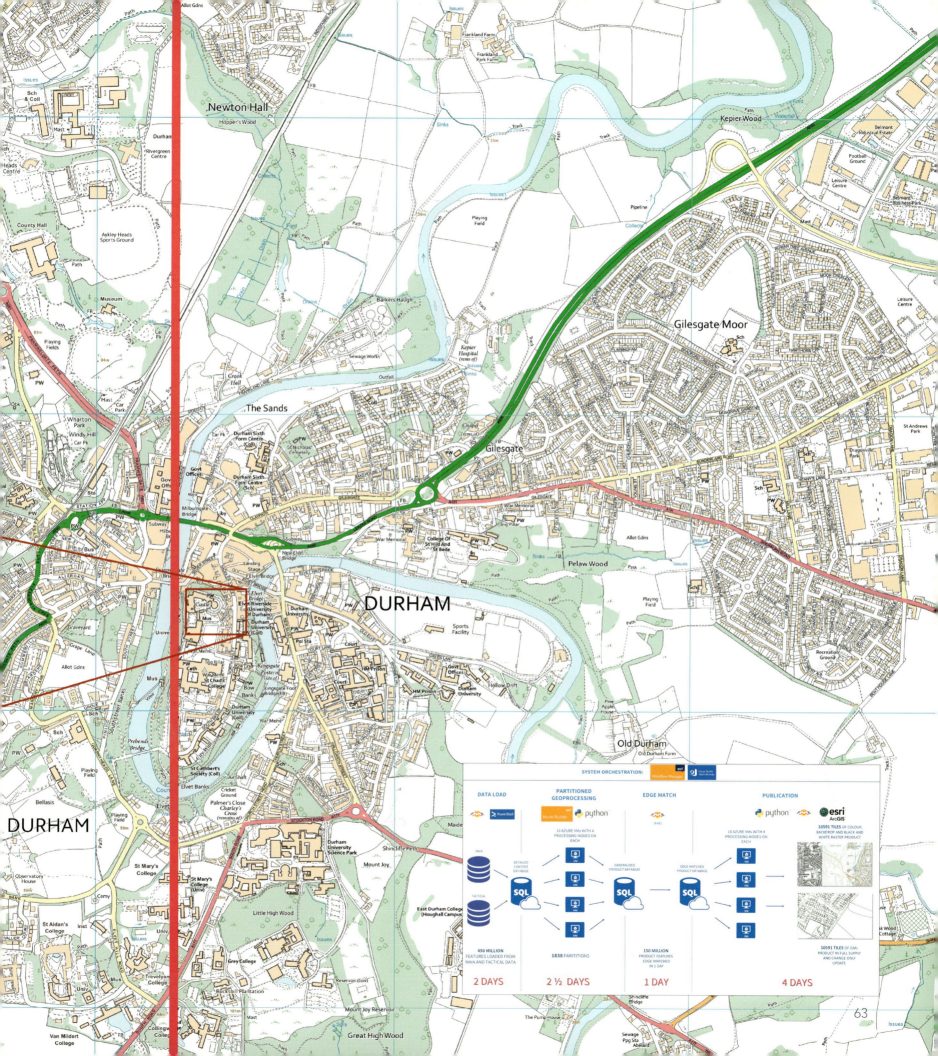

FROM COUNTRY TO CITY

Cache County
Logan, Utah, USA
By Jacob Adams

Urban growth and development are a fact of modern life in Cache Valley. Situated just over the mountains from the booming Wasatch Front and home to a leading engineering and agricultural university, demand continues to grow for new homes and businesses in this once-rural community. If we want to conserve some of the valley's agricultural character in the process, we first need to identify the patterns of development.

By assigning structures a different color based on the decade they were built with a single group for all existing structures built before 1950, we can see how subdivisions have spread out from the historical city grids laid out by the early pioneers. Many cities are now separated only by an imaginary line on a map, and large-lot agricultural subdivisions are encroaching on rural land far from established cities.

CONTACT

Jacob Adams
jacob.dan.adams@gmail.com

SOFTWARE

ArcGIS Pro, Microsoft® Excel, GIMP

DATA SOURCES

Cache County Development Services, Cache County Assessor, State of Utah State Geographic Information Database, US Geological Survey, US Census Bureau, US Department of Agriculture Farm Service Agency

Courtesy of Cache County Development Services Department.

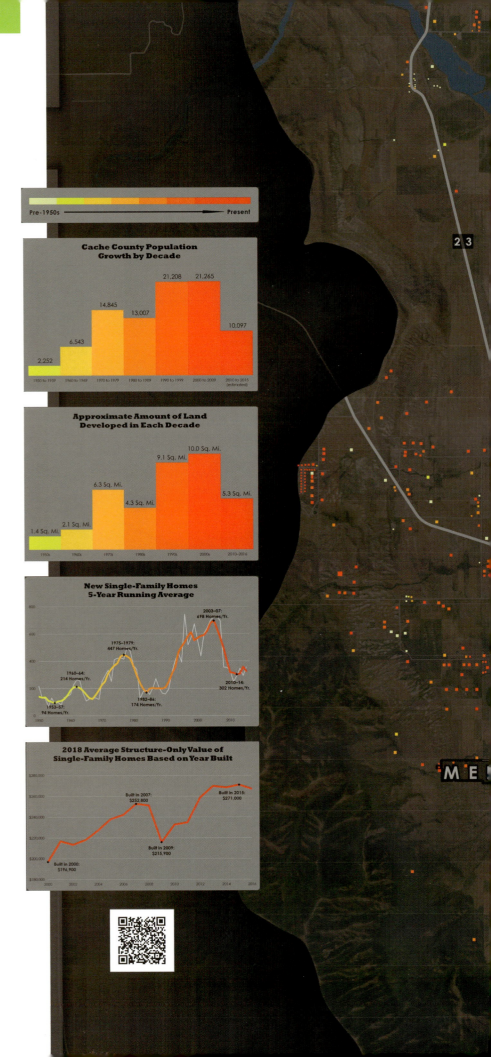

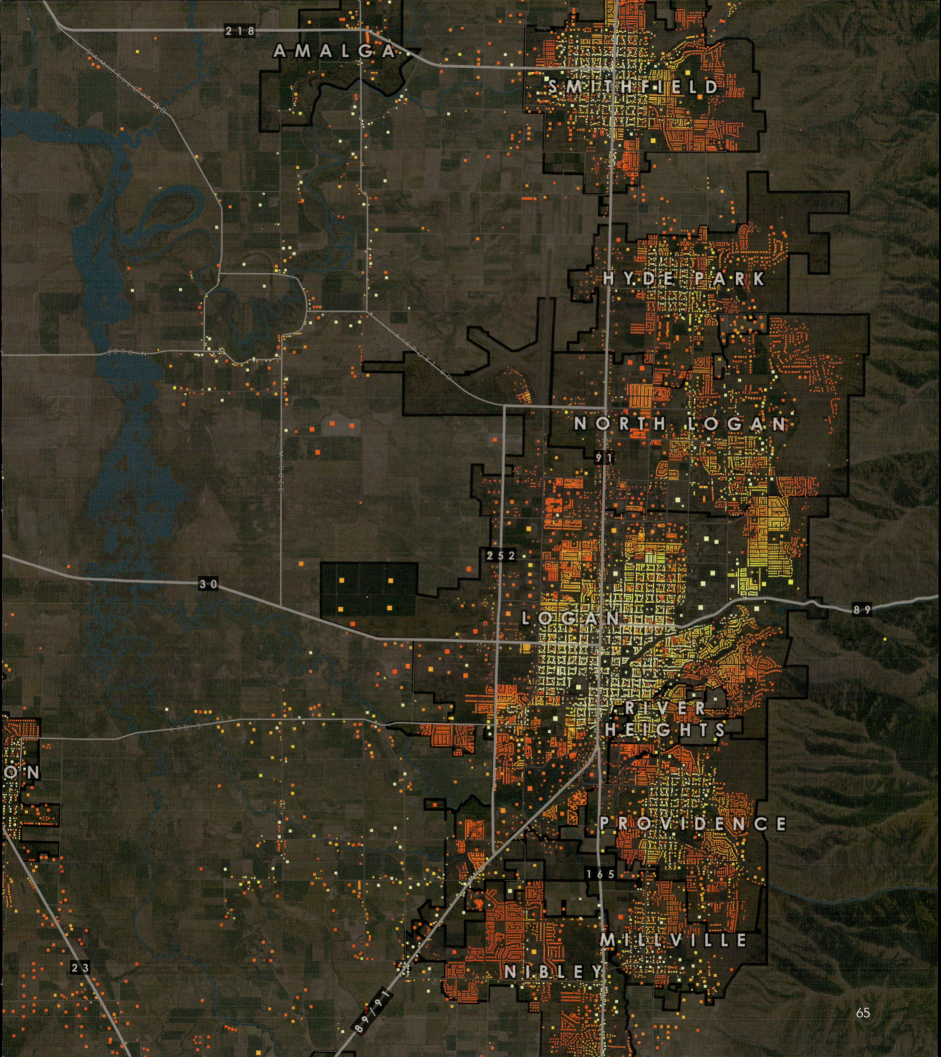

AMALGA

218

SMITHFIELD

HYDE PARK

NORTH LOGAN

91

252

LOGAN

RIVER
HEIGHTS

30

89

600 S

PROVIDENCE

165

MILLVILLE

NIBLEY

23

89/91

65

CLIMATE-SMART CITIES™: DENVER

Green infrastructure strategies for climate equity and resilience

The Trust for Public Land
Santa Fe, New Mexico, USA
By Emily Patterson, Lara Miller, Kristen Weil, Danny Paez, Lindsey Rotche, and Lindsay Withers

Denver is highly vulnerable to the impacts of a changing climate and is projected to face warmer days, changing snow pack, more intense wildfires, and more intense storms and precipitation events over the next several decades. With a growing economy and a densifying city, strategic action is needed now for Denver to develop equitable and sustainable solutions to expected climate impacts.

The Trust for Public Land's national Climate-Smart Cities™ program is providing key planning and decision-making support to help the city and county of Denver take strategic action on climate change through green infrastructure and prepare for a climate-resilient future with a particular emphasis on investments for under-served populations.

CONTACT
Lara Miller
lara.miller@tpl.org

SOFTWARE
ArcGIS Desktop 10.5, ModelBuilder

DATA SOURCES
The Trust for Public Land, Esri, City and County of Denver

Copyright © The Trust for Public Land. Information on these maps is provided for purposes of discussion and visualization only.

Urban heat islands
Climate-Smart Cities™ Denver: Cool Priorities

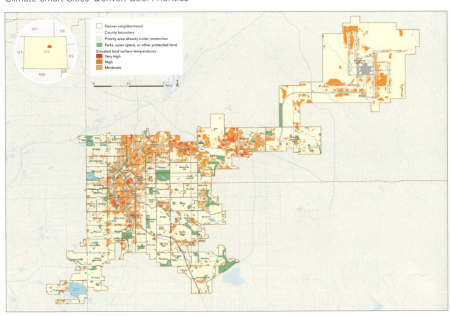

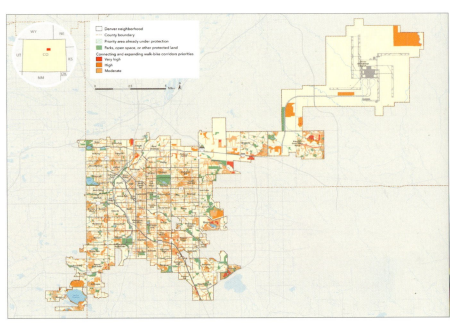

Connecting and expanding walk-bike corridors
Climate-Smart Cities™ Denver: Connect Priorities

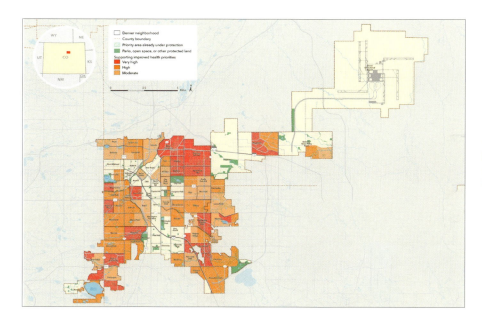

Supporting improved
health for all residents
Climate-Smart Cities™ Denver:
Health Equity Priorities

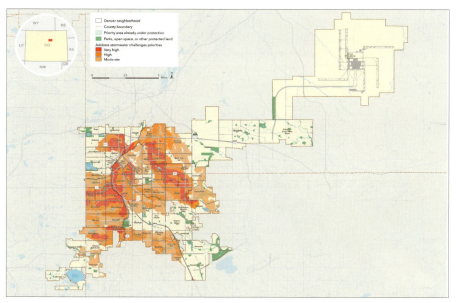

Addressing stormwater
and flooding challenges
Climate-Smart Cities™ Denver:
Absorb and Protect Priorities

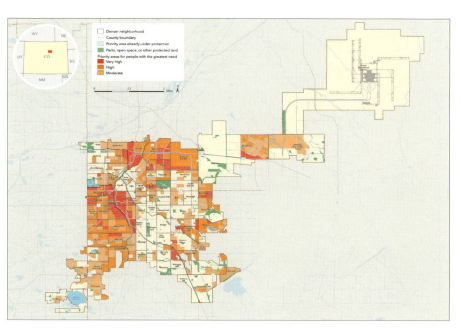

Focusing on people with
the greatest need
Climate-Smart Cities™ Denver:
Climate Equity Priorities

HOW "THE WEST'S MOST WESTERN TOWN" HAS GROWN

City of Scottsdale
Scottsdale, Arizona, USA
By Mele D. Koneya

Scottsdale, Arizona, is known as a vacation and business destination that began as a small farming community with a rich history of ranches, farms, and settlers. On June 25, 1951, Scottsdale officially adopted "The West's Most Western Town" as its motto to reflect its small-town, western heritage.

The town expanded rapidly during the 1950s, growing to a population of more than 10,000 within an area of about five square miles by 1960. By the end of the 1960s, Scottsdale's population had increased six-fold to nearly 68,000 while its land area increased twelve-fold to 62 square miles.

The following decades brought even more growth in population and land area. By 1980, its population of more than 88,000 covered 88.6 square miles. By 1990, it had reached more than 130,000 in population and expanded to roughly its present size—about 185 square miles. Today, the city is home to more than 230,000 residents.

This map was created to show Scottsdale's growth by construction year of the assessed parcels.

CONTACT
Mele D. Koneya
mkoneya@scottsdaleaz.gov

SOFTWARE
ArcGIS Pro 2.2, Adobe InDesign®

DATA SOURCES
City of Scottsdale, Maricopa County Assessor, Scottsdale Public Library, Scottsdale Historical Society

Courtesy of City of Scottsdale GIS.

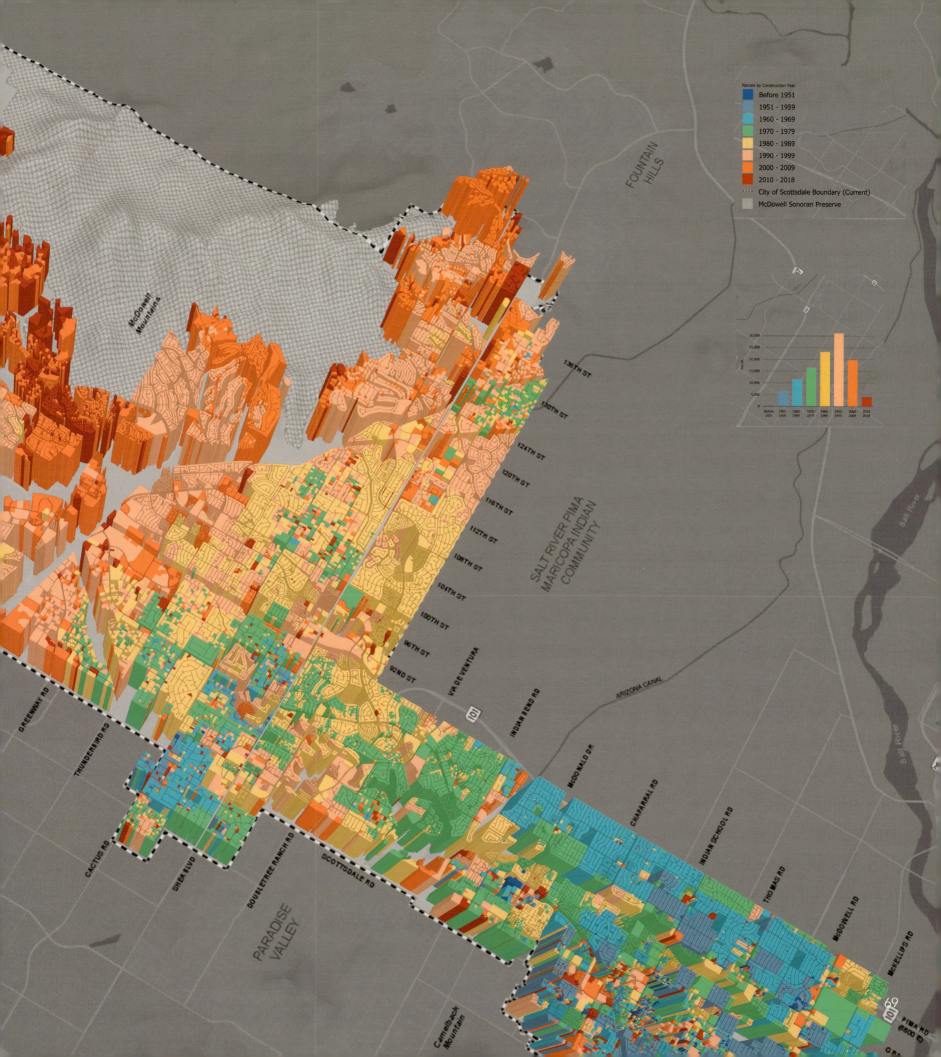

Parcels by Construction Year

- Before 1951
- 1951 - 1959
- 1960 - 1969
- 1970 - 1979
- 1980 - 1989
- 1990 - 1999
- 2000 - 2009
- 2010 - 2018
- ······ City of Scottsdale Boundary (Current)
- McDowell Sonoran Preserve

FOUNTAIN HILLS

McDowell Mountains

SALT RIVER PIMA MARICOPA INDIAN COMMUNITY

Salt River

ARIZONA CANAL

PARADISE VALLEY

Camelback Mountain

GREENWAY RD

THUNDERBIRD RD

CACTUS RD

SHEA BLVD

DOUBLETREE RANCH RD

SCOTTSDALE RD

VIA DE VENTURA

92ND ST

96TH ST

100TH ST

104TH ST

108TH ST

112TH ST

116TH ST

120TH ST

124TH ST

130TH ST

136TH ST

INDIAN BEND RD

MCDONALD DR

CHAPARRAL RD

INDIAN SCHOOL RD

THOMAS RD

MCDOWELL RD

MCKELLIPS RD

PIMA RD

101

101

IMPLEMENTATION OF ARCGIS FOR PORTAL AND ARCGIS ONLINE BY VICKSBURG DISTRICT, USACE

US Army Corps of Engineers-Vicksburg District
Vicksburg, Mississippi, USA
By Brian Everitt, Julie Vicars, Bill Sisneros, and Kim Day

With the introduction of Esri's suite of enterprise geospatial technologies, Vicksburg District has implemented several web applications for data creation and management as well as data sharing across multiple offices, partners, and organizations. The benefits of this technology have been the establishment of a common operating picture allowing for a more efficient means of gathering, collating, synthesizing, and disseminating information to all appropriate parties involved, no matter their location or their area of expertise. Vicksburg District geospatial professionals have applied this robust technology across multiple subject areas, ranging from internal emergency management applications to public-facing recreation applications. With the dynamic, ever-changing nature of this technology, functionality and application is limitless.

CONTACT
Brian Everitt
Brian.C.Everitt@usace.army.mil

SOFTWARE
ArcGIS Online, ArcPortal 10.6, ArcGIS Desktop 10.4.1, ArcGIS Pro 2.0

DATA SOURCES
US Geological Survey, National Oceanic and Atmospheric Administration, Federal Emergency Management Agency, Department of Homeland Secutiry, National Weather Service, US Army Corps of Engineers

Courtesy of US Army Corps of Engineers-Vicksburg District.

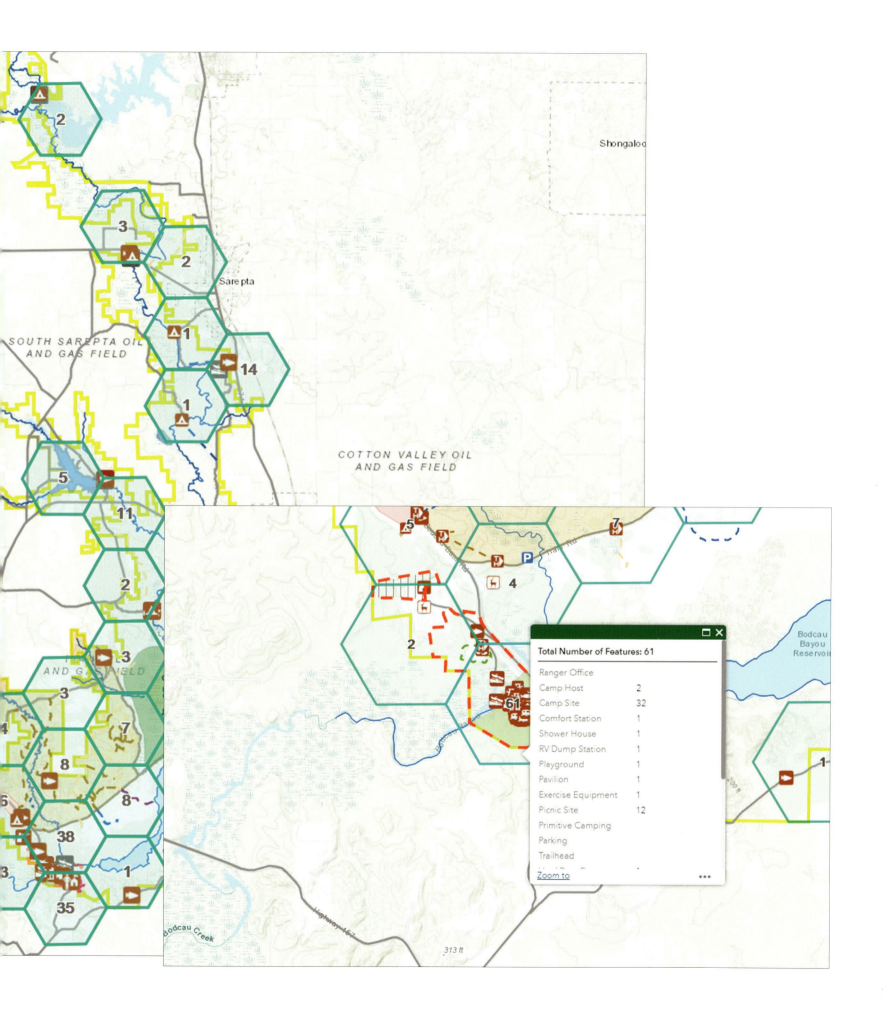

Shongaloo

SOUTH SAREPTA OIL
AND GAS FIELD

Sarepta

COTTON VALLEY OIL
AND GAS FIELD

AND GAS FIELD

Bodcau Creek

Bodcau
Bayou
Reservoir

313 ft

Total Number of Features: 61

Ranger Office	
Camp Host	2
Camp Site	32
Comfort Station	1
Shower House	1
RV Dump Station	1
Playground	1
Pavilion	1
Exercise Equipment	1
Picnic Site	12
Primitive Camping	
Parking	
Trailhead	

Zoom to

UNDERSTANDING MARYVALE

Maricopa Association of Governments
Phoenix, Arizona, USA
By Deborah Brown, Jami Dennis, Mark Roberts, and
Christian Gort

The Maricopa Association of Governments (MAG) is a
council of governments (COG) that serves as the regional
planning agency for the metropolitan Phoenix area. MAG
provides regional planning and policy decisions in areas of
transportation, air quality, water quality, and human services.

This analysis shows an overview of socioeconomic factors
that provide background information for revitalization
efforts within Maryvale Village, which is the most populous
of the fifteen urban villages in Phoenix, Arizona. The result is
a general overview of the demographics, income, housing,
education, and economy of Maryvale Village. The analysis
also shows a comparison of socioeconomic data with the
larger city of Phoenix and Maricopa County areas that
encompass Maryvale Village.

CONTACT
Deborah Brown
dbrown@azmag.gov

SOFTWARE
ArcGIS Desktop 10.5, Microsoft® Publisher 2016,
Microsoft Excel 2016

DATA SOURCES
US Census Bureau, MAG, Maricopa County, Arizona
Department of Education, The Information Market,
Maricopa County Transportation Reduction Program

Courtesy of Maricopa Association of Governments.

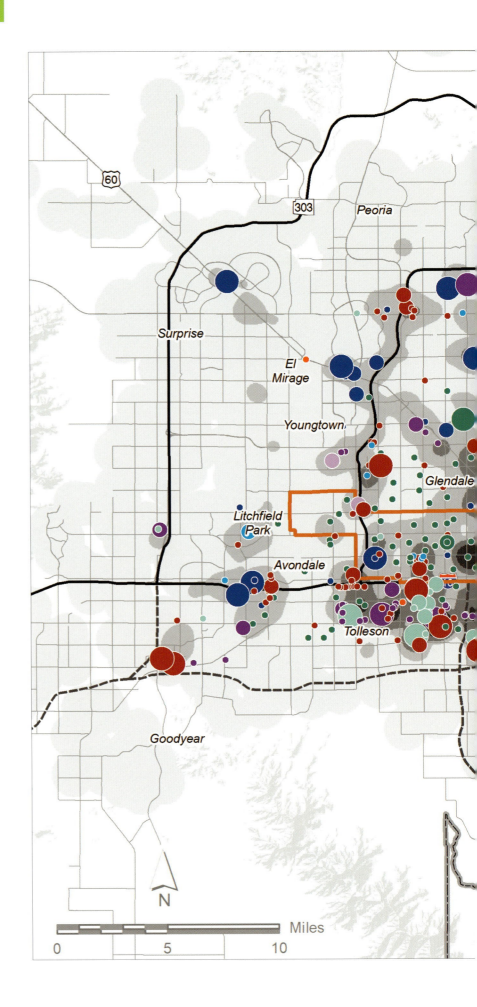

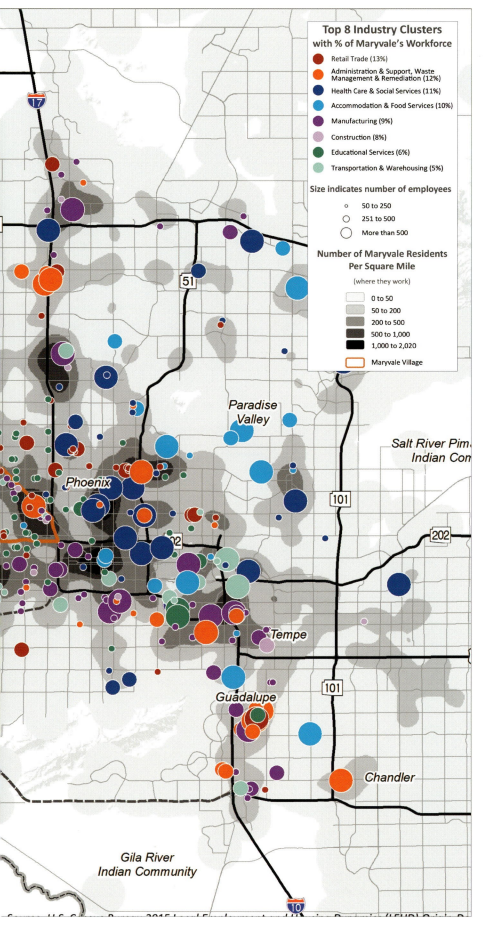

Top 8 Industry Clusters
with % of Maryvale's Workforce

- ● Retail Trade (13%)
- ● Administration & Support, Waste Management & Remediation (12%)
- ● Health Care & Social Services (11%)
- ● Accommodation & Food Services (10%)
- ● Manufacturing (9%)
- ● Construction (8%)
- ● Educational Services (6%)
- ● Transportation & Warehousing (5%)

Size indicates number of employees

- ○ 50 to 250
- ○ 251 to 500
- ○ More than 500

Number of Maryvale Residents Per Square Mile

(where they work)

- 0 to 50
- 50 to 200
- 200 to 500
- 500 to 1,000
- 1,000 to 2,020
- ☐ Maryvale Village

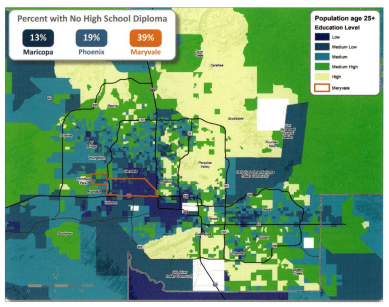

Percent with No High School Diploma

13%	19%	39%
Maricopa	Phoenix	Maryvale

Population age 25+ Education Level
- Low
- Medium Low
- Medium
- Medium High
- High
- ☐ Maryvale

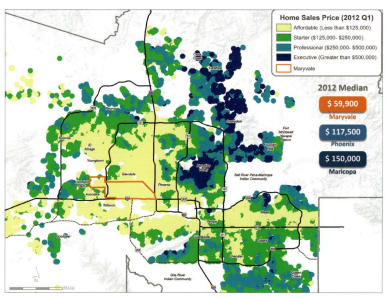

Home Sales Price (2012 Q1)
- Affordable (Less than $125,000)
- Starter ($125,000-$250,000)
- Professional ($250,000-$500,000)
- Executive (Greater than $500,000)
- ☐ Maryvale

2012 Median

$ 59,900	Maryvale
$ 117,500	Phoenix
$ 150,000	Maricopa

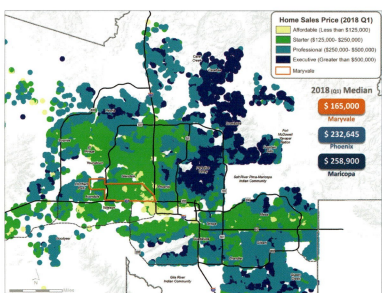

Home Sales Price (2018 Q1)
- Affordable (Less than $125,000)
- Starter ($125,000-$250,000)
- Professional ($250,000-$500,000)
- Executive (Greater than $500,000)
- ☐ Maryvale

2018 (Q1) Median

$ 165,000	Maryvale
$ 232,645	Phoenix
$ 258,900	Maricopa

ANALYSIS OF THE ADDRESSES TO POLLING SITES

Elections Canada (Electoral Geography Division)
Gatineau, Quebec, Canada
By Elections Canada

This map displays an analysis of addresses to polling sites for the Federal electoral district of Ottawa–Vanier. This proximity analysis pilot project consists of determining if electors are assigned to the polling site that is the closest to their home address using polling site elector capacity numbers and the road network with the help of the Network Analyst extension in ArcGIS. The areas where there are addresses of concern, electors not assigned to the optimal polling site, are displayed in red. The proximity analysis methodology was later refined and national distance thresholds were established to flag the addresses of concern where electors' residences required them to travel distances that are above the thresholds to get to their assigned polling site. These cases will be reviewed by the local Electoral Administrators when they execute their pre-electoral activities in order to make voting more accessible to electors. Statistics are generated in this project to draw conclusions from the data.

CONTACT
Francisco Tomas
francisco.tomas@elections.ca

SOFTWARE
ArcGIS Desktop 10.4.1

DATA SOURCES
Elections Canada (Electoral Geography Division), Natural Resources Canada (GeoAccess, Canada Centre for Remote Sensing) and Statistics Canada (Geography Division).

Courtesy of Elections Canada.

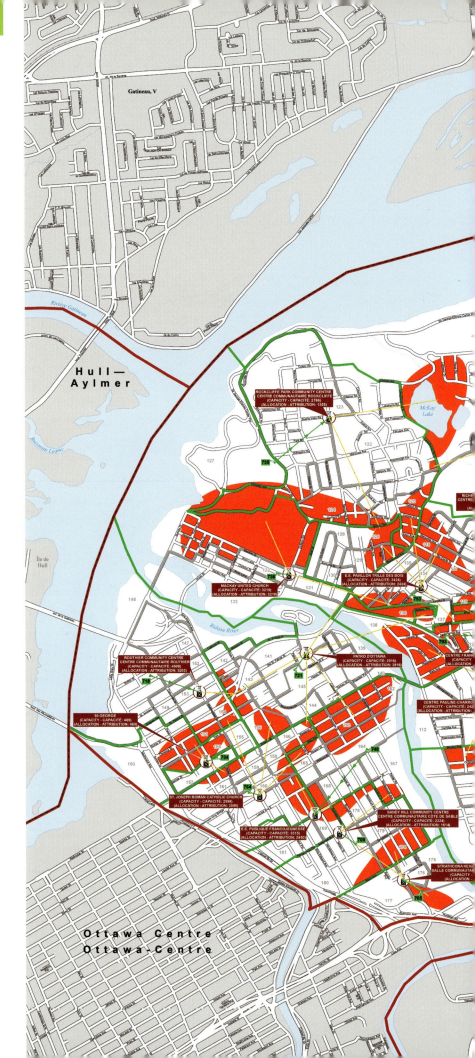

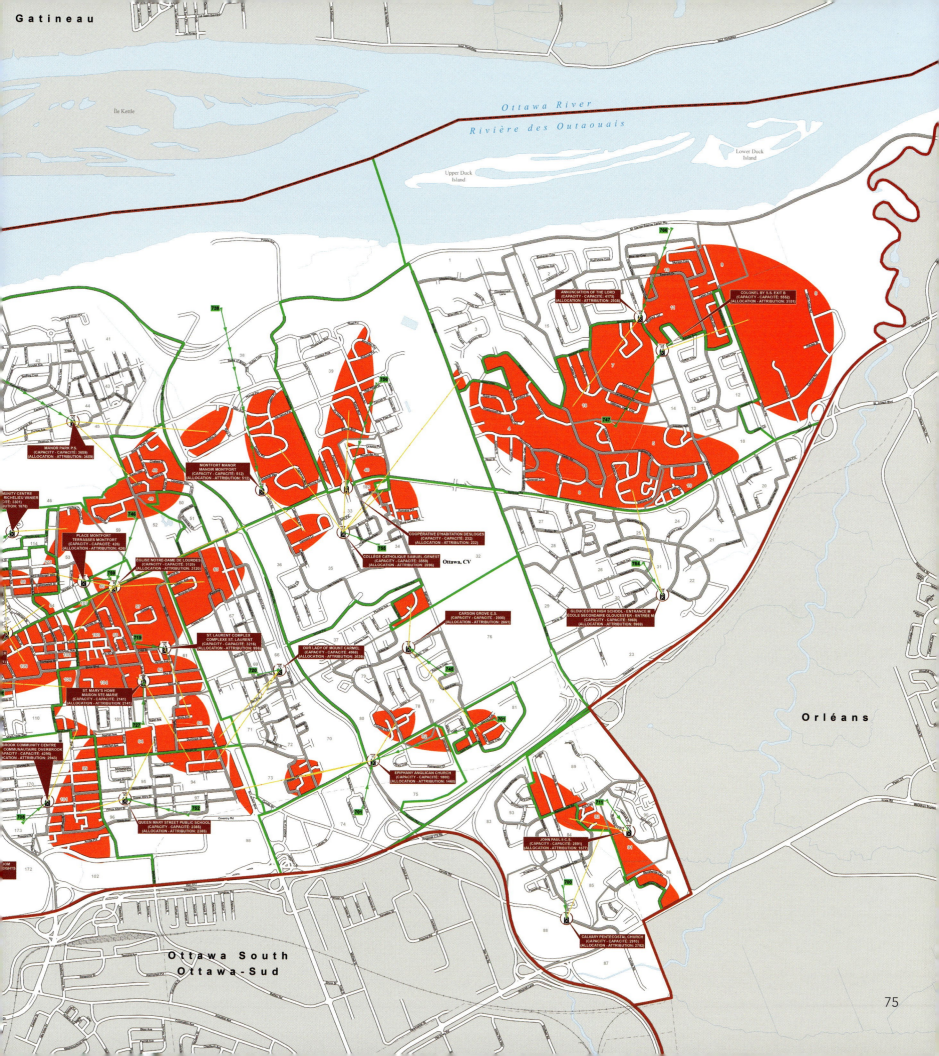

Gatineau

Ottawa River
Rivière des Outaouais

Île Kettle

Lower Duck
Island

Upper Duck
Island

Orléans

ANNUNCIATION OF THE LORD
(CAPACITY - CAPACITÉ: 4170)
(ALLOCATION - ATTRIBUTION: 2938)

COLONEL BY S.S. EXIT B
(CAPACITY - CAPACITÉ: 5552)
(ALLOCATION - ATTRIBUTION: 3101)

MANOR PARK P.S.
(CAPACITY - CAPACITÉ: 3659)
(ALLOCATION - ATTRIBUTION: 3659)

MONTFORT MANOR
MANOIR MONTFORT
(CAPACITY - CAPACITÉ: 512)
(ALLOCATION - ATTRIBUTION: 512)

MUNITY CENTRE
RICHELIEU VANIER
ITÉ: 3301)
UTION: 1878)

PLACE MONTFORT
TERRASSES MONTFORT
(CAPACITY - CAPACITÉ: 426)
(ALLOCATION - ATTRIBUTION: 426)

ÉGLISE NOTRE-DAME DE LOURDES
(CAPACITY - CAPACITÉ: 3120)
(ALLOCATION - ATTRIBUTION: 3120)

COOPÉRATIVE D'HABITATION DESLOGES
(CAPACITY - CAPACITÉ: 232)
(ALLOCATION - ATTRIBUTION: 232)

COLLÈGE CATHOLIQUE SAMUEL-GENEST
(CAPACITY - CAPACITÉ: 3859)
(ALLOCATION - ATTRIBUTION: 2896)

Ottawa, CV

CARSON GROVE E.S.
(CAPACITY - CAPACITÉ: 2990)
(ALLOCATION - ATTRIBUTION: 2861)

GLOUCESTER HIGH SCHOOL - ENTRANCE M
ÉCOLE SECONDAIRE GLOUCESTER - ENTRÉE M
(CAPACITY - CAPACITÉ: 5969)
(ALLOCATION - ATTRIBUTION: 5969)

ST. LAURENT COMPLEX
COMPLEXE ST. LAURENT
(CAPACITY - CAPACITÉ: 3216)
(ALLOCATION - ATTRIBUTION: 956)

OUR LADY OF MOUNT CARMEL
(CAPACITY - CAPACITÉ: 4960)
(ALLOCATION - ATTRIBUTION: 3038)

ST. MARY'S HOME
MAISON STE-MARIE
(CAPACITY - CAPACITÉ: 2141)
(ALLOCATION - ATTRIBUTION: 2141)

BROOK COMMUNITY CENTRE
COMMUNAUTAIRE OVERBROOK
(CAPACITY - CAPACITÉ: 4295)
(ATION - ATTRIBUTION: 2943)

EPIPHANY ANGLICAN CHURCH
(CAPACITY - CAPACITÉ: 1900)
(ALLOCATION - ATTRIBUTION: 1460)

QUEEN MARY STREET PUBLIC SCHOOL
(CAPACITY - CAPACITÉ: 2385)
(ALLOCATION - ATTRIBUTION: 2385)

JOHN PAUL II C.E.
(CAPACITY - CAPACITÉ: 2591)
(ALLOCATION - ATTRIBUTION: 1677)

CALVARY PENTECOSTAL CHURCH
(CAPACITY - CAPACITÉ: 2591)
(ALLOCATION - ATTRIBUTION: 2782)

Ottawa South
Ottawa-Sud

75

SPATIAL ASSESSMENT OF HIGH-RISK LEAD POISONING AREAS IN LOS ANGELES COUNTY

Los Angeles County
Los Angeles, California, USA
By Douglas Morales

The US Department of Housing and Urban Development (HUD) Lead Hazard Reduction Demonstration (LHRD) Grant "funds units of state, local and tribal government to implement comprehensive programs to identify and remediate lead-based paint hazards in privately owned rental or owner-occupied housing." Lead poisoning remains a public health concern in Los Angeles County. Therefore, GIS was used to create a composite index to spatially assess high-risk areas of lead poisoning in order to identify a study area for the HUD LHRD grant. The following Los Angeles County census tract data was used in the analyses: pre-1940 housing; children below the age of six; poverty measures; and elevated blood lead levels. This data was recoded, scored, ranked, combined, and visualized based on the HUD LHRD scoring criteria. Consequently, HUD awarded $3.4 million to Los Angeles County to reduce the number of children with elevated blood lead levels and protect families living in homes with significant lead and other home health and safety hazards.

(Disclaimer: The methodology used in this project was based on the HUD LHRD scoring criteria and was not affiliated with the State of California Childhood Lead Poisoning Prevention Branch.)

CONTACT
Douglas Morales
dmorales@ph.lacounty.gov

SOFTWARE
ArcGIS Desktop 10.5.1

DATA SOURCES
County of Los Angeles Childhood Lead Poisoning Prevention Program, US Census Bureau

Courtesy of Los Angeles County.

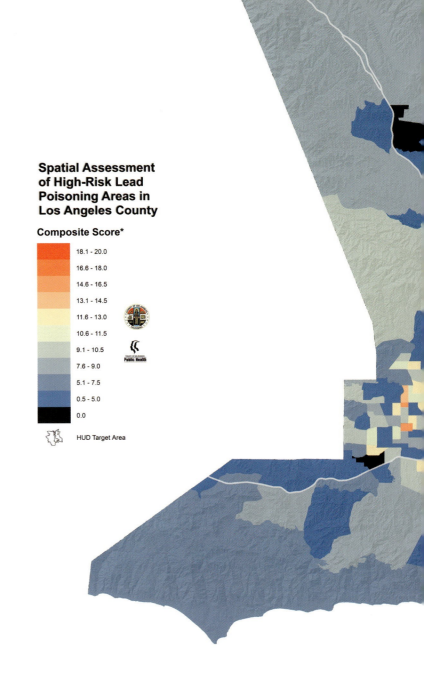

Spatial Assessment of High-Risk Lead Poisoning Areas in Los Angeles County

Composite Score*

- 18.1 - 20.0
- 16.6 - 18.0
- 14.6 - 16.5
- 13.1 - 14.5
- 11.6 - 13.0
- 10.6 - 11.5
- 9.1 - 10.5
- 7.6 - 9.0
- 5.1 - 7.5
- 0.5 - 5.0
- 0.0

HUD Target Area

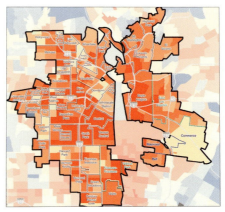

Labels:
Los Angeles City Communities = Black
Incorporated Cities = Blue
Unincorporated Areas = Grey

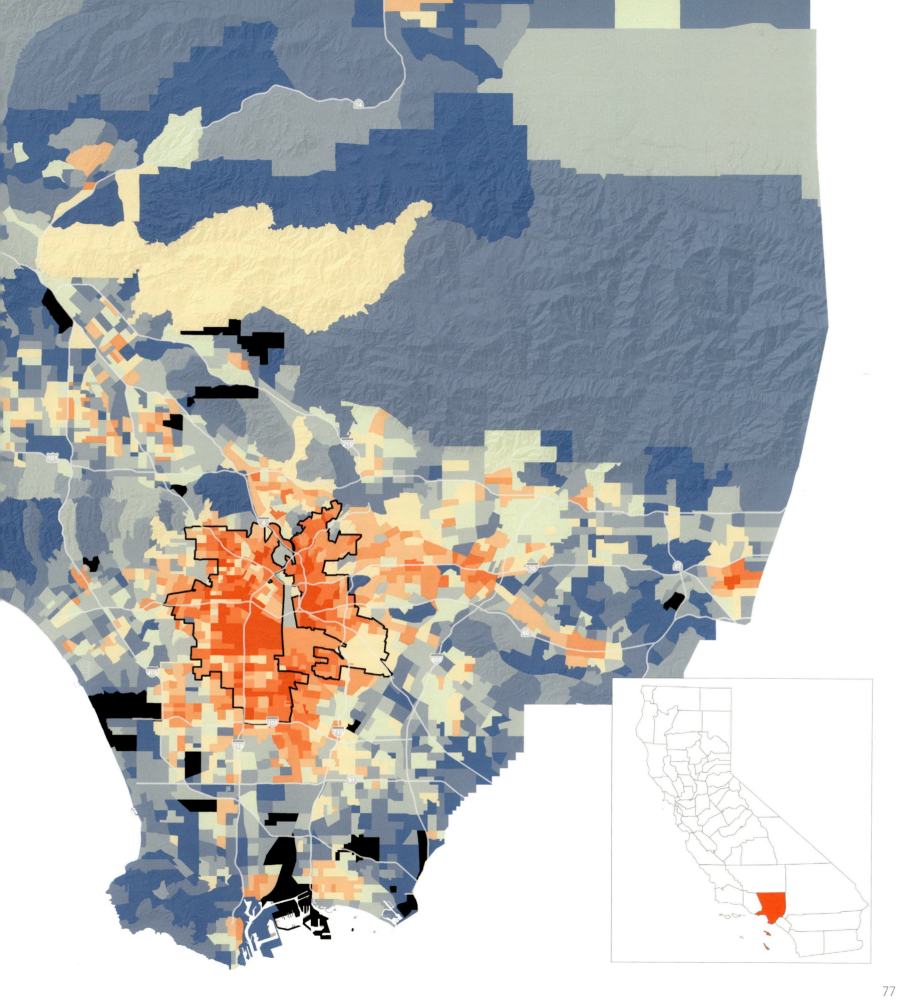

A GEOSPATIAL VISUALIZATION TOOL TO ASSESS ZIKA VIRUS TRANSMISSION RISK UNDER ALTERNATIVE SCENARIOS

Abt Associates
Boulder, Colorado, USA
By Russell Jones, Michelle Krasnec, and Ryan Takeshita

Zika virus (ZIKV) has recently been recognized as a significant threat to global public health. Although ZIKV is present in large parts of the Western Hemisphere, little is known about factors affecting the transmission of ZIKV. Infected travelers have contributed to the spread of ZIKV in countries where suitable mosquitoes have become infected and propagated local transmission cycles, demonstrating the importance of understanding patterns of human movement in the global spread of ZIKV.

We used a combination of data from laboratory infectivity studies with mosquitos, current and projected environmental data related to mosquito source populations, and human travel patterns to conduct a geospatial examination to assess the threat of further ZIKV outbreaks in the US. We present the geospatial risk of ZIKV spread using several illustrative transmission scenarios. Our results indicate that the probability of ZIKV outbreaks largely depends on several factors affecting transmission, including the mosquito species and strains that are competent vectors, and the climatic conditions suitable for their survival.

We describe geospatial tools that can help decision makers assess baseline risk for local ZIKV transmission and assist them with making efficient and timely decisions regarding vector control activities. These tools can also help decision makers target the most threatening Zika-competent mosquito subpopulations for a given area.

CONTACT

Russell Jones
russ_jones@abtassoc.com

SOFTWARE

ArcGIS Desktop 10.4

DATA SOURCES

Airport Data: Bureau of Transportation Stats; Base Climate: PRISM Climate Group, Oregon State University; Future Climate: CLIMsystems Ltd., 2017; Mosquito data: Kraemer MUG, et. al., 2015

Courtesy of Abt Associates.

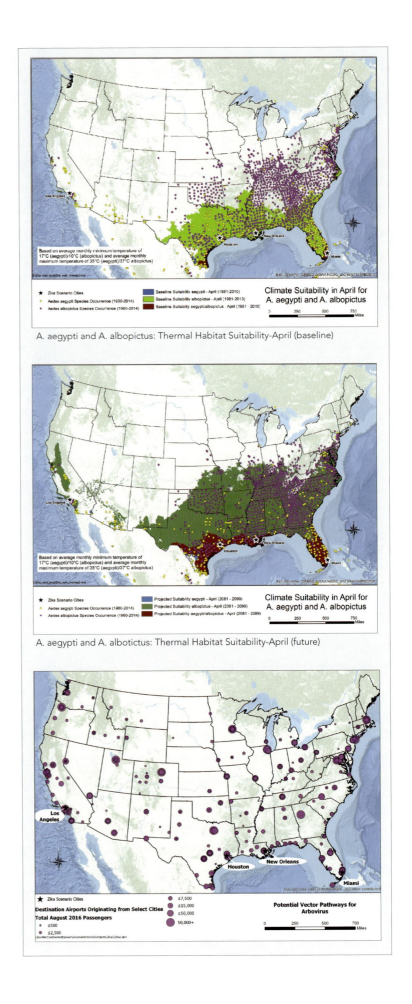

A. aegypti and A. albopictus: Thermal Habitat Suitability-April (baseline)

A. aegypti and A. albotictus: Thermal Habitat Suitability-April (future)

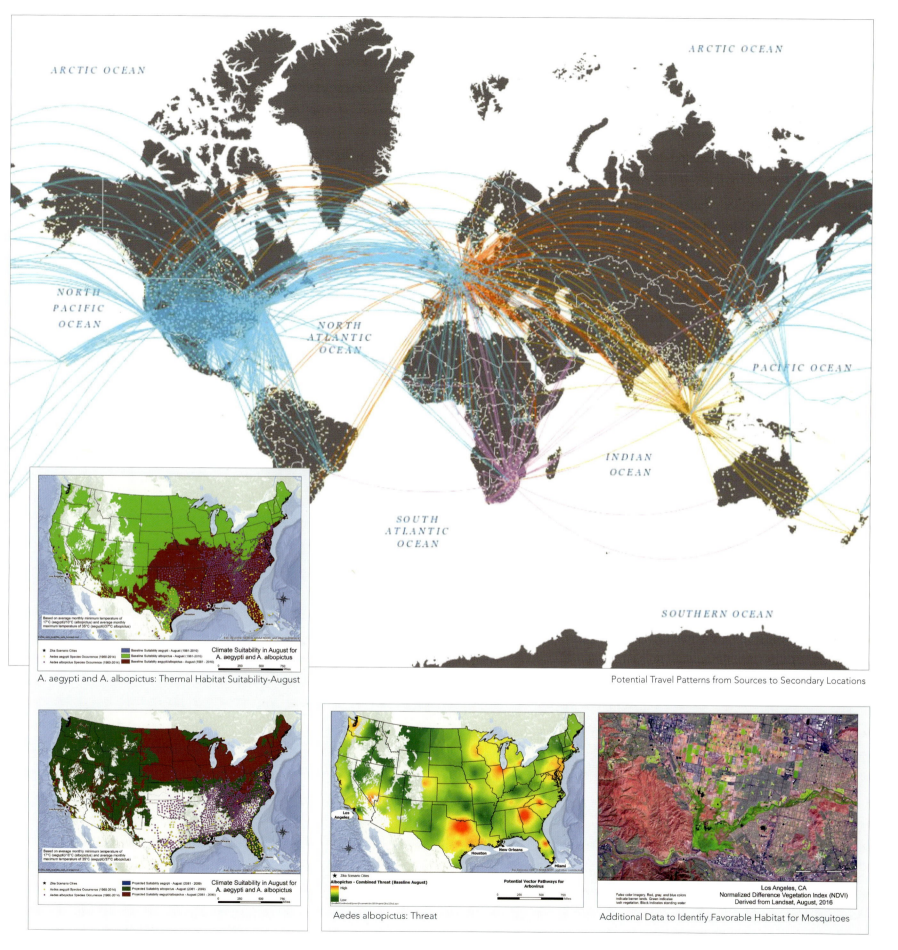

A. aegypti and A. albopictus: Thermal Habitat Suitability-August

Potential Travel Patterns from Sources to Secondary Locations

Aedes albopictus: Threat

Additional Data to Identify Favorable Habitat for Mosquitoes

IMPROVING ACCESS TO CARE– TANANA CHIEFS CONFERENCE HEALTH SERVICES MASTER PLANNING EFFORT

The Innova Group
Tucson, Arizona, USA
By Elizabeth Ayarbe-Perez, John Temple, Anthony Laird, Carmen Zamora, and Kent Tarbet

Healthcare delivery is population based. The greater the population, the greater the demand. Therefore, understanding population growth is an essential element of the healthcare planning task.

This mapping effort was part of a Health Services Master Plan for the Tanana Chiefs Conference (TCC) in Alaska. The goal of this project was to understand TCC's future comprehensive demand for healthcare services, accessibility options, and capital investment requirements supported by implementation priorities.

It is critical to map the physical accessibility of care. TCC serves 235,000 square miles of remote interior Alaska where access to care is challenging. These maps strengthen the planner's ability to discover the geographic challenges while observing population and workload differences between communities.

Clearly articulating how the present network of care functions at village, sub-regional, and regional locations supports appropriate future service area definitions, sustainable on-site health services, natural referral patterns, and improved accessibility to care, resulting in implementable priorities and healthier populations.

CONTACT
Elizabeth Ayarbe-Perez
elizabeth.perez@theinnovagroup.com

SOFTWARE
ArcGIS Pro 2.0

DATA SOURCES
TCC GIS, OpenFlights, Esri basemaps

Courtesy of The Innova Group.

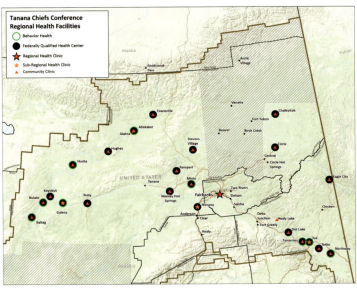

Present Network of Care

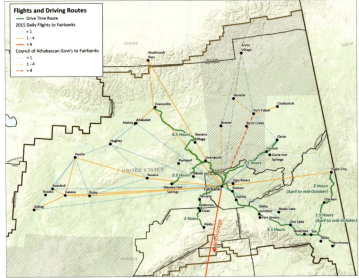

Access Opportunities

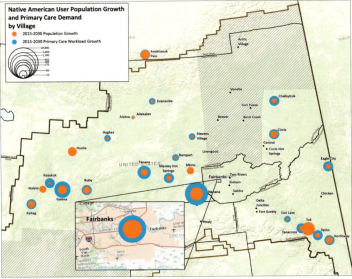

Future Demand

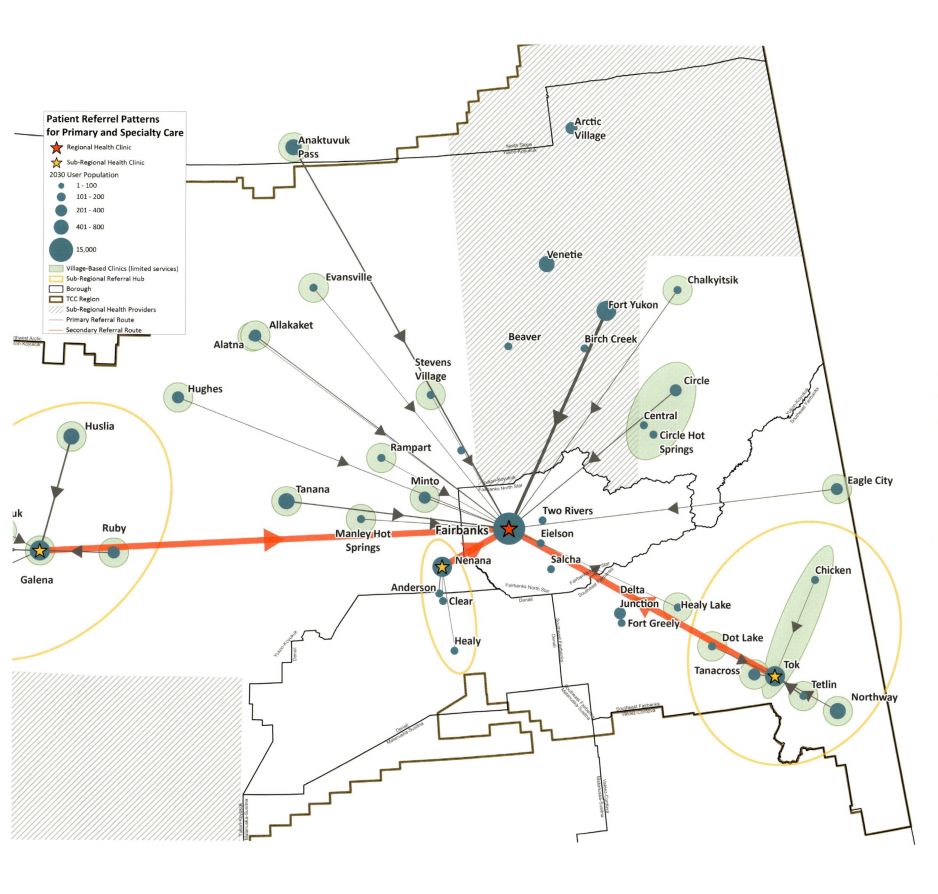

Patient Referrel Patterns for Primary and Specialty Care

⬟ Regional Health Clinic
⬟ Sub-Regional Health Clinic

2030 User Population
- 1 - 100
- 101 - 200
- 201 - 400
- 401 - 800
- 15,000

Village-Based Clinics (limited services)
Sub-Regional Referral Hub
Borough
TCC Region
Sub-Regional Health Providers
Primary Referral Route
Secondary Referral Route

Anaktuvuk Pass

Arctic Village

Venetie

Chalkyitsik

Evansville

Fort Yukon

Allakaket

Beaver

Birch Creek

Alatna

Stevens Village

Circle

Hughes

Central

Circle Hot Springs

Huslia

Eagle City

Rampart

Tanana

Minto

Two Rivers

Ruby

Manley Hot Springs

Fairbanks

Eielson

Galena

Nenana

Salcha

Anderson

Clear

Delta Junction

Healy Lake

Chicken

Fort Greely

Healy

Dot Lake

Tanacross

Tok

Tetlin

Northway

North Slope
Yukon-Koyukuk

Northwest Arctic
Yukon-Koyukuk

Yukon-Koyukuk
Southeast Fairbanks

Yukon-Koyukuk
Fairbanks North Star

Fairbanks North Star
Southeast Fairbanks

Fairbanks North Star
Denali

Denali
Southeast Fairbanks

Yukon-Koyukuk
Denali

Denali
Matanuska-Susitna

Denali
Matanuska-Susitna

Southeast Fairbanks
Matanuska-Susitna

Southeast Fairbanks
Valdez-Cordova

Denali
Matanuska-Susitna

TUBERCULOSIS AND DIABETES MELLITUS IN LAC, 2015-2017

Los Angeles County
Los Angeles, California, USA
By Monica Rosales, PhD, MS; Edward Lan, MPH; Josephine Yumul, MSc; Ramon E. Guevara, PhD, MPH; Douglas M. Morales, MPH; Julie M. Higashi, MD, PhD; and Alicia H. Chang, MD, MS

GIS methods are important epidemiological tools for identifying and assessing geographic variations in disease morbidity. In Los Angeles County (LAC), the rate of active tuberculosis (TB) disease is higher than the state and national rates. This project examined the geographic distribution of the co-occurrence of TB and diabetes mellitus (DM) in LAC by identifying areas and populations disproportionately burdened.

Since LAC is comprised of twenty-four health districts that are used to plan and manage health service delivery, distribution of TB and DM burden was assessed at the health district level. These maps highlight hot spots encompassing several health districts among all TB cases, all TB cases with DM, Asian TB cases with DM, and Hispanic TB cases with DM. These findings highlight opportunities for community-based outreach initiatives in areas at high risk for disease burden and for targeted public health interventions aimed at reducing health disparities in underserved communities.

CONTACT
Edward Lan
elan@ph.lacounty.gov

SOFTWARE
ArcGIS Desktop 10.3.1

DATA SOURCES
Los Angeles County Department of Public Health
Tuberculosis Registry Information Management System

Courtesy of Los Angeles County Department of Public Health Tuberculosis Control Program.

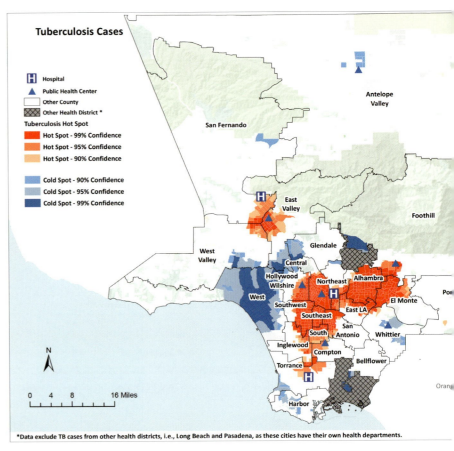

*Data exclude TB cases from other health districts, i.e., Long Beach and Pasadena, as these cities have their own health departments.

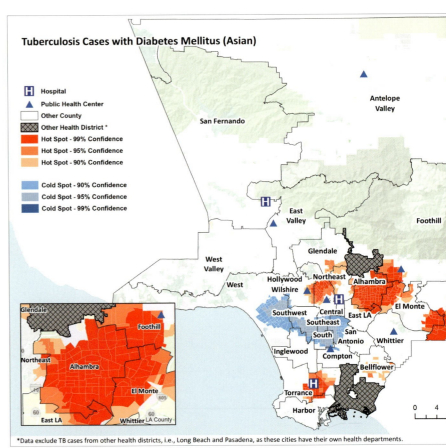

*Data exclude TB cases from other health districts, i.e., Long Beach and Pasadena, as these cities have their own health departments.

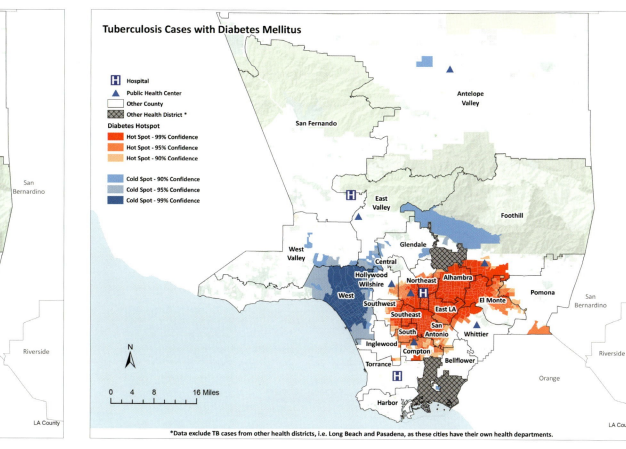

Tuberculosis Cases with Diabetes Mellitus

Legend:
- **H** Hospital
- ▲ Public Health Center
- Other County
- Other Health District *

Diabetes Hotspot
- Hot Spot - 99% Confidence
- Hot Spot - 95% Confidence
- Hot Spot - 90% Confidence

- Cold Spot - 90% Confidence
- Cold Spot - 95% Confidence
- Cold Spot - 99% Confidence

*Data exclude TB cases from other health districts, i.e. Long Beach and Pasadena, as these cities have their own health departments.

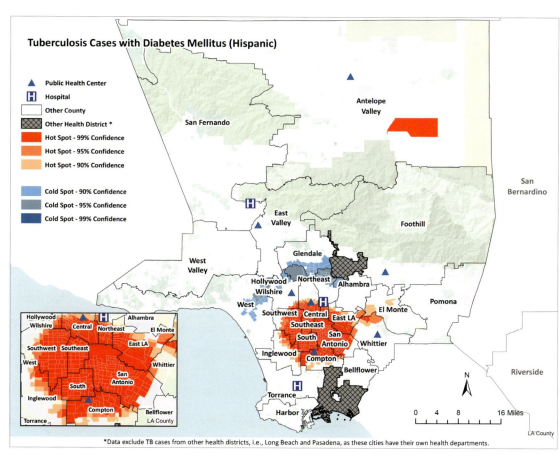

Tuberculosis Cases with Diabetes Mellitus (Hispanic)

Legend:
- ▲ Public Health Center
- **H** Hospital
- Other County
- Other Health District *
- Hot Spot - 99% Confidence
- Hot Spot - 95% Confidence
- Hot Spot - 90% Confidence

- Cold Spot - 90% Confidence
- Cold Spot - 95% Confidence
- Cold Spot - 99% Confidence

*Data exclude TB cases from other health districts, i.e., Long Beach and Pasadena, as these cities have their own health departments.

NOT IF BUT WHEN: A RISK ANALYSIS FOR RABIES ENTERING AUSTRALIA

Tufts University
Taylorsville, Utah, USA
By Jenny Schilling

Rabies virus is a fatal disease that can be transmitted between mammals and humans through contact with an infected animal's saliva, typically through a bite. Although rabies is preventable through vaccines, this virus remains endemic to much of the world, with some countries experiencing an increase in incidence of rabies in recent years. However, Australia is unique in its historical ability for successful prevention of the intrusion of rabies within its borders, accomplished through extremely strict import, quarantine, and vaccine protocols involving domestic animals.

Unfortunately, these strict precautions may soon be in vain as rabies continues to run rampant in countries surrounding Australia, notably the Indonesian archipelago and Papua New Guinea, both of which participate in heavy trade with Australia and are within 300 kilometers off the northern coast of the continent. Rabies can be contracted by any of the 300+ wild mammal species that inhabit Australia, spreading between domestic animals, livestock, and humans. This is especially risky for wide-ranging endemic carnivore species such as the wild dingo and red fox, whom are ideal carries and transmitters of the virus. The inherent and inevitable risk of rabies entering Australia could be less damaging if there is an understanding of the potential entry points, disease spread, and factors that play an important role in mitigating this deadly virus.

CONTACT
Jenny Schilling
jennifer.schilling@tufts.edu

SOFTWARE
ArcGIS Desktop 10.5.1

DATA SOURCES
National Center for Ecological Analysis and Synthesis (NCEAS), National Exposure Information Systems (NEXIS), International Union for Conservation of Nature (IUCN), Australian Department of the Environment, Data.gov.au

Courtesy of Tufts University.

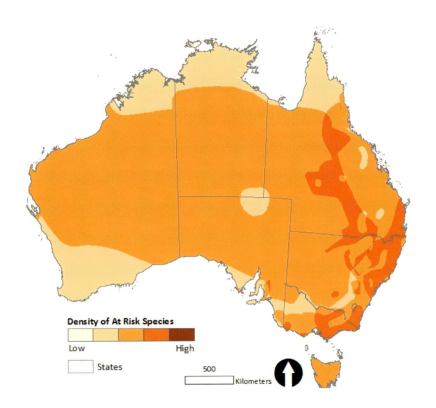

Density of At Risk Species

Low · High

States · 500 Kilometers

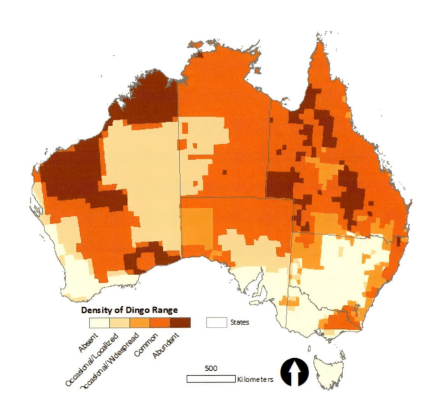

Density of Dingo Range

Absent · Occasional/Localized · Occasional/Widespread · Common · Abundant · States

500 Kilometers

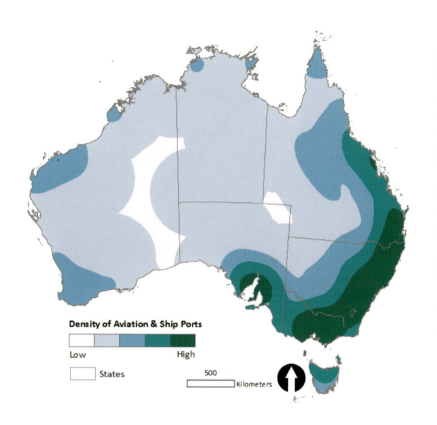

Density of Aviation & Ship Ports

Low ▭▭▭▭▭ High

▢ States

500 Kilometers

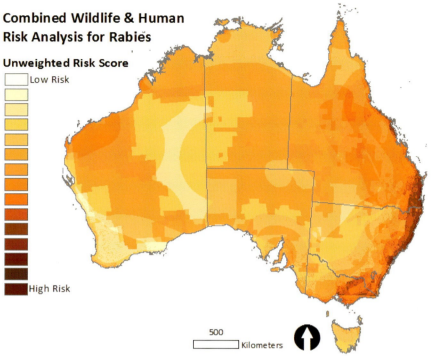

Combined Wildlife & Human Risk Analysis for Rabies

Unweighted Risk Score

Low Risk

High Risk

500 Kilometers

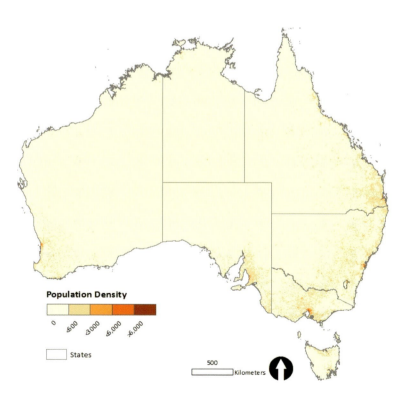

Population Density

0   <500   <3,000   <6,000   >6,000

▢ States

500 Kilometers

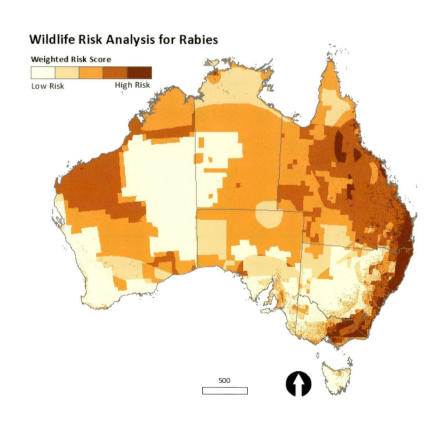

Wildlife Risk Analysis for Rabies

Weighted Risk Score

Low Risk   High Risk

500

MEDIEVAL CASTLE DOCUMENTATION USING THE METHODS OF GIS

University of Jan Evangelista Purkyne, Faculty of Environment
Usti nad Labem, Czech Republic
By Jan Pacina

This map shows the result of documenting Rýzmburk, a medieval castle located in central Europe, using the methods of geoinformatics. Detailed photogrammetric survey, laser scanning, and field survey were performed in order to document the current state of this important part of our cultural heritage. The processed data available online in the form of vector layers (contour lines, walls, and building structures), 3D models, and photo plans of the various preserved structures are used for documentation of the castle reconstruction, scientific (archaeological) research, and general tourist information. On the presented map, you may see a detailed orthophoto (spatial resolution 5 cm/pixel) showing the main castle and the surroundings together with elevation information and topographical features. The fortification of the castle (three large towers and walls) is clearly visible on the orthophoto. The orthophoto was created using the small-format aerial photography method. The different points on the map represent the created 3D models that can be opened by clicking on them. Here, we can find detailed 3D models of the castle structures (red triangles) and 3D models of archaeological soundings (blue circles).

CONTACT
Jan Pacina
jan.pacina@ujep.cz

SOFTWARE
ArcGIS Desktop 10.4, ArcGIS Online, Agisoft PhotoScan 3.0

DATA SOURCES
University of Jan Evangelista Purkyně Faculty of Environment

Courtesy of FZP UJEP.

ARCHEOACOUSTICS OF THE ANASAZI LANDSCAPE

Kristy E. Primeau
Sand Lake, New York, USA
By Kristy E. Primeau

For acoustic archaeologists, putting sound back into the archaeological landscape is an important part of understanding how people lived, what they valued, how they shaped their identities, and how they experienced the world and their place in it. This soundshed map shows how sound produced by a conch shell trumpet would have spread between great houses, shrines, and other archaeological site locations within the Anasazi landscape more than a thousand years ago. The sound of the conch can be seen travelling to a number of locations that often marked canyon overlooks or high points on the landscape.

The human experience of sound is studied by mapping Sound Pressure Levels, measured in decibels (dB). Musical instruments, such as Anasazi conch shell trumpets, have been linked to ritual performance. By studying what the Anasazi heard, archaeologists can understand where people began to actively participate in and experience a ritual occurring at ceremonial structures, like Casa Rincoñada, and other important sites.

CONTACT
Kristy Primeau
kprimeau@albany.edu

SOFTWARE
ArcGIS Desktop 10.3

DATA SOURCES
Primeau and Witt 2018 (https://doi.org/10.1016/j.jasrep.2017.05.044), National Park Service

This map has been adapted from Primeau and Witt (2018) and from Primeau and Witt (2017).

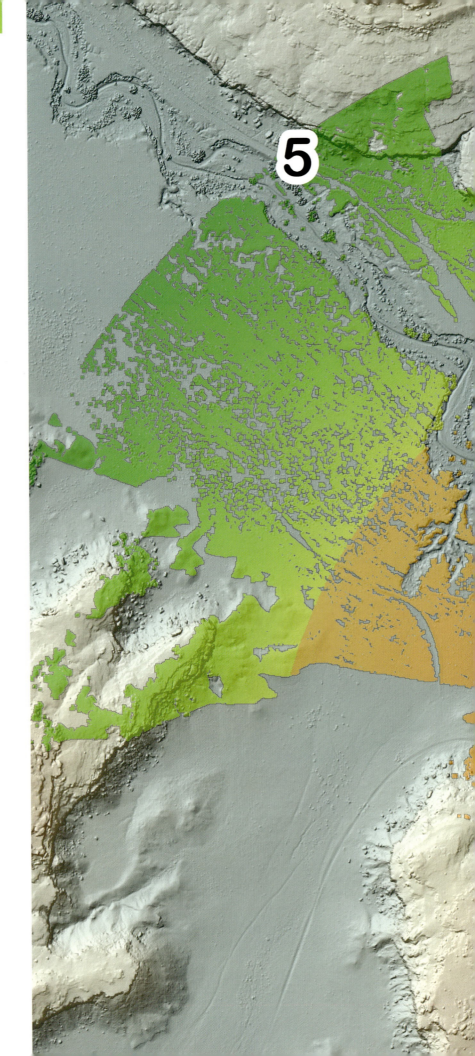

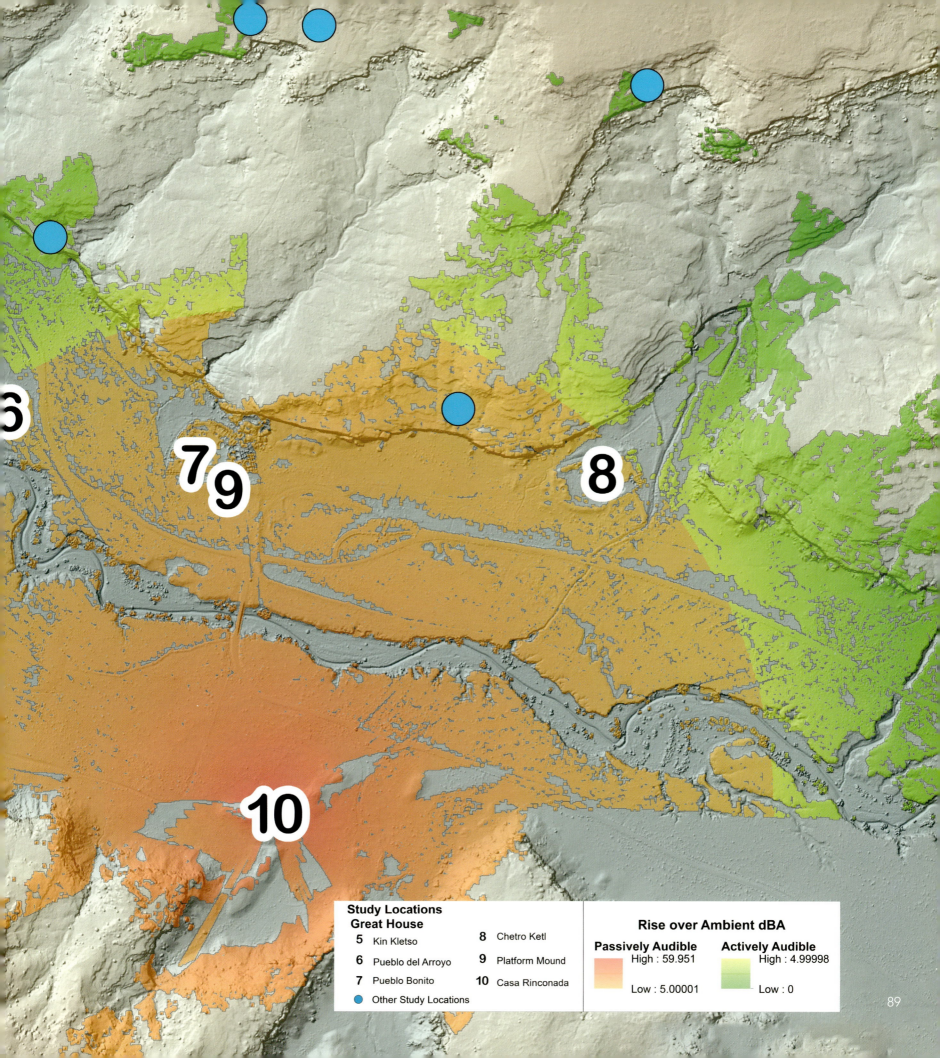

Study Locations
Great House

5 Kin Kletso

6 Pueblo del Arroyo

7 Pueblo Bonito

8 Chetro Ketl

9 Platform Mound

10 Casa Rinconada

● Other Study Locations

Rise over Ambient dBA

Passively Audible
High : 59.951

Low : 5.00001

Actively Audible
High : 4.99998

Low : 0

UNEMPLOYMENT RATES

PCBS
Ramallah, Palestine
By Nafeer Massad

This map was produced by the Palestinian Central Bureau of Statistics in order to help researchers and decision makers by highlighting a crucial issue: unemployment rates among Palestinian youth graduates between eighteen-to-twenty-nine years old with two years diploma and above. This was 47.6 percent with a high discrepancy at the regional level, between 26.6 percent in the West Bank and 69.2 percent in Gaza Strip.

The data was taken from the third Population, Housing and Establishments Census 2017, which was carried out using tablets and GIS tools for the first time. New technology is adopted to keep abreast of the technological development in the field of data collection, as well as to save time, effort, and cost; reduce time between data collection; shorten dissemination; and to increase the quality and accuracy of data.

CONTACT
Nafeer Massad
nafir@pcbs.gov.ps

SOFTWARE
ArcGIS Desktop 10.2.2

DATA SOURCES
Palestine Population, Housing and Establishment Census, 2017

Courtesy of PCBS, 2018.

	0.0 - 7.1
	7.2 - 20.3
	20.4 - 30.1
	30.2 - 43.3
	43.4 - 64.9
	65.0 - 100.0
	Unavailable Data
	Israeli Settlements
	Expansion & Annexation Wall

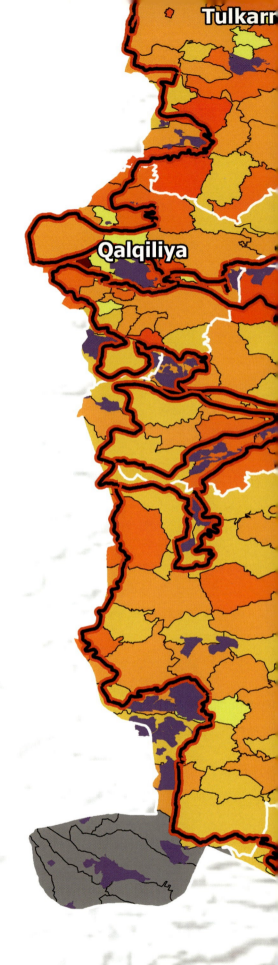

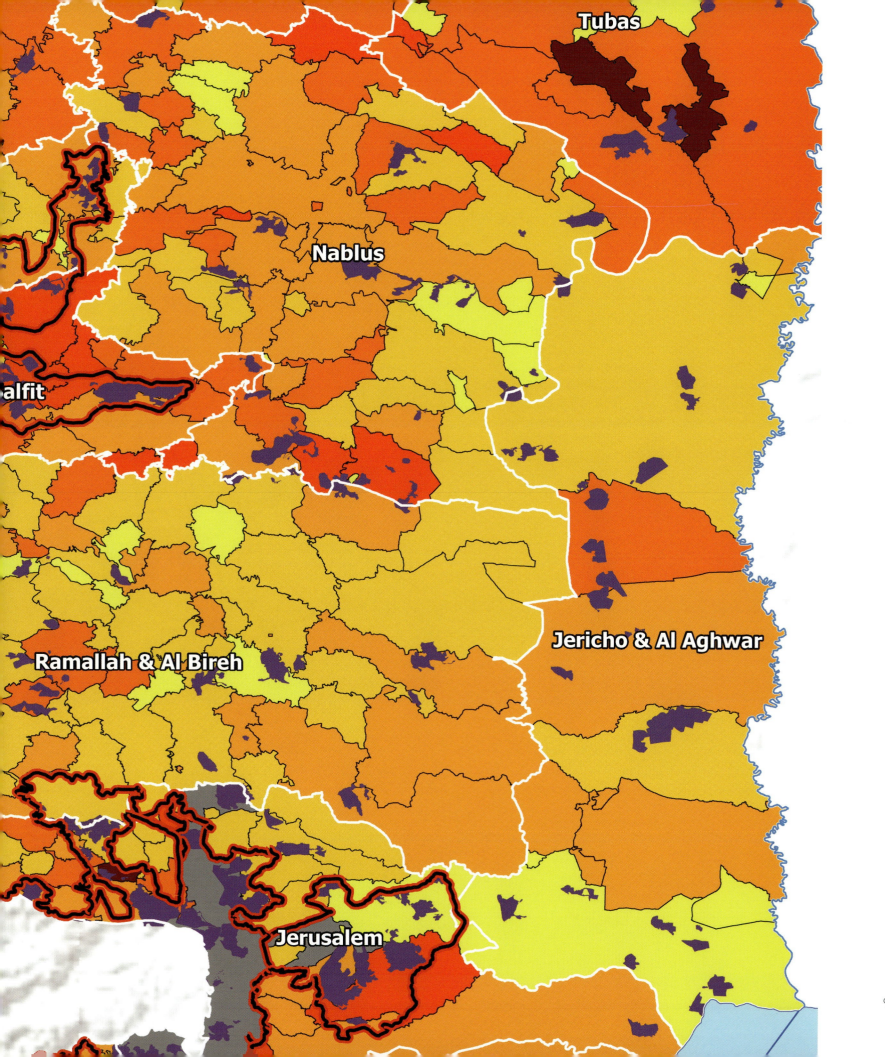

ORDER OF MAGNITUDE OF DEBRIS PRODUCED IN MOSUL AFTER ISIL CONTROL AND LIBERATION OPERATIONS

UN-Habitat Iraq Programme
Baghdad, Iraq
By Karrar Al-Eqabi

A preliminary debris assessment of Mosul is now online. This initial mapping estimates that there is almost 11 million tons of residential debris as a result of the destruction in the city. This is a preliminary estimation that should be regarded as an absolute minimum and is subject to verification in the field. The actual amount may be several factors higher, due to the expected voluntary demolition of severely damaged buildings, among others. More detailed investigations indicate that the debris volume is in the range of approximately 7.6 million tons.

This debris map and assessment was produced by United Nations Environment Programme (UNEP), Urban Resilience Platform (URP), and United Nations for Human Settlement Programme (UN-Habitat).

CONTACT
Karrar Al-Eqabi
karrar.yousif@un.org

SOFTWARE
ArcGIS Desktop 10.5, Adobe Illustrator

DATA SOURCES
Disaster Waste Recovery, URP, UNEP, UN-Habitat

Courtesy of UN-Habitat Iraq Programme.

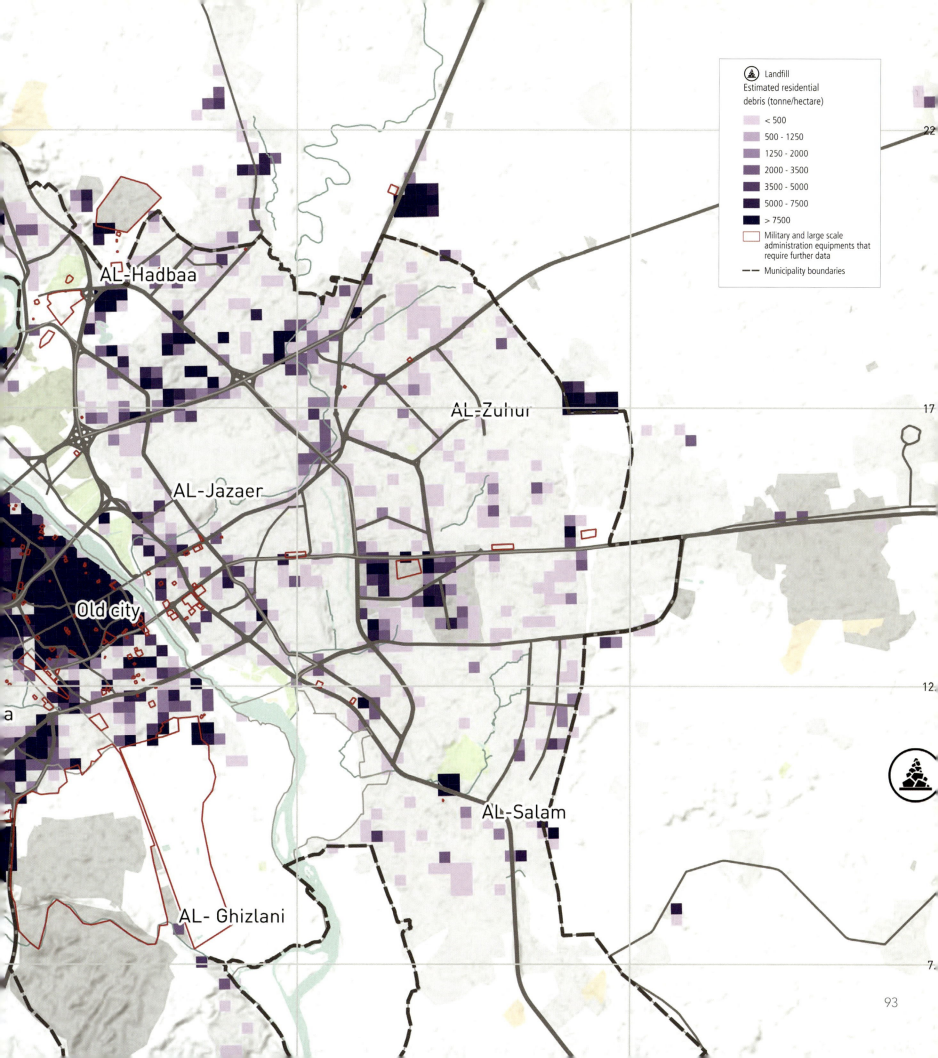

AL-Hadbaa

AL-Zuhur

AL-Jazaer

Old city

a

AL-Salam

AL- Ghizlani

Landfill
Estimated residential
debris (tonne/hectare)

< 500

500 - 1250

1250 - 2000

2000 - 3500

3500 - 5000

5000 - 7500

> 7500

Military and large scale
administration equipments that
require further data

Municipality boundaries

93

ISRAEL CENTER DISTRICT AGRICULTURE MAPPING

Ministry of Agriculture and Rural Development
Rishon Lezion, Center, Israel
By Omer Ben-Asher

These maps present the diversity of Israeli agriculture that succeeds to develop despite the difficult environmental conditions. The mapping project carried out in the Central District of the Ministry of Agriculture and Rural Development in the last decade shows the agricultural branches, different types of water sources used for irrigation, different cover interfaces that improve the production efficiency, and the characteristics of the unique settlement forms.

CONTACT
Omer Ben-Asher
omerb@moag.gov.il

SOFTWARE
ArcGIS Desktop

DATA SOURCES
A survey of the agricultural settlements and interpretation of aerial photographs

Courtesy of Israel Center District and Ministry of Agriculture and Rural Development, Israel.

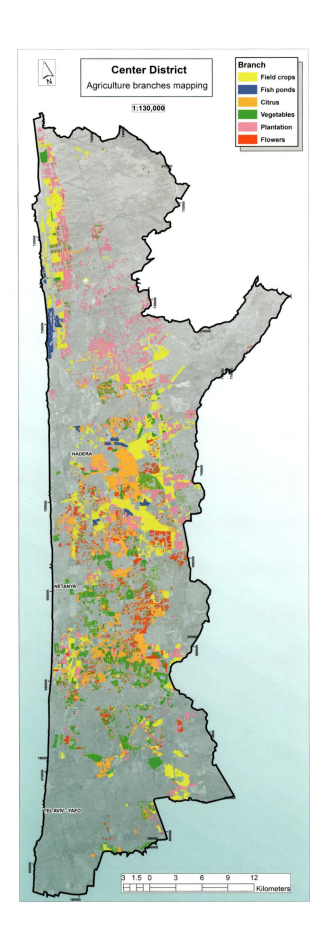

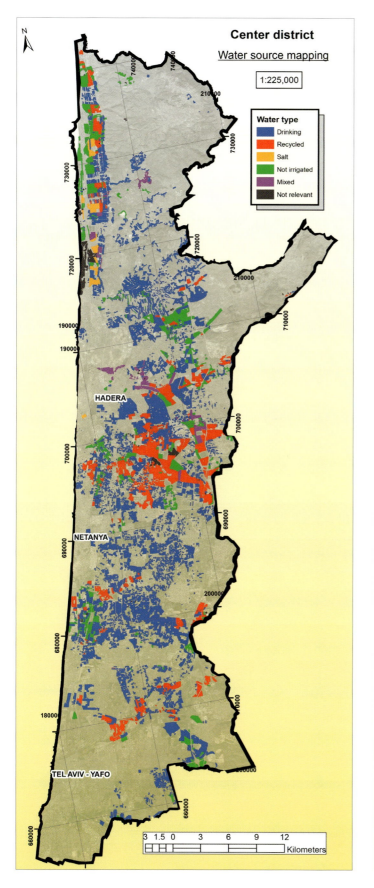

Center district

Water source mapping

1:225,000

Water type
- Drinking
- Recycled
- Salt
- Not irrigated
- Mixed
- Not relevant

HADERA

NETANYA

TEL AVIV - YAFO

3 1.5 0 3 6 9 12 Kilometers

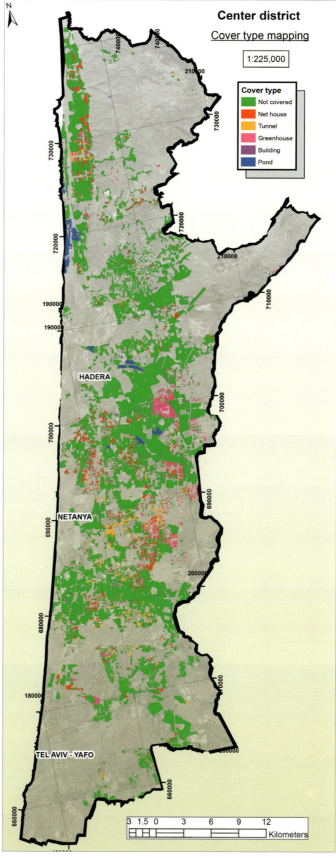

Center district

Cover type mapping

1:225,000

Cover type
- Not covered
- Net house
- Tunnel
- Greenhouse
- Building
- Pond

HADERA

NETANYA

TEL AVIV - YAFO

3 1.5 0 3 6 9 12 Kilometers

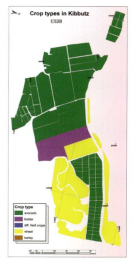

Crop types in Kibbutz

Crop type
- avocado
- fodder
- diff field crops
- wheat
- barley

Crop types in Moshav

crop type
- banana
- banana in net house
- diff. field crops
- olive
- eggplant
- cauliflower
- not cultivated
- lemon
- Cucumber in greenhouse
- trees
- tomatoes in greenhouse
- tomatoes in an open area
- pepper in greenhouse
- pepper high technology
- pepper for export
- flowers

OPERATIONAL FLOOD MONITORING OF AGRICULTURE DURING HURRICANES HARVEY AND IRMA WITH RADAR DATA

USDA/NASS
Washington, District of Columbia, USA
By Claire G. Boryan, Zhengwei Yang, Avery Sandborn, and Patrick Willis

Agricultural flood monitoring is important for food security and economic stability and is of significant interest to the United States Department of Agriculture's National Agricultural Statistics Service (NASS). In agricultural remote sensing applications, optical sensor data are traditionally used for acreage, yield, and crop condition assessments. However, optical data is affected by cloud cover, rain, and darkness. These limitations restrict the capability of optical data to assess flood disaster events in a timely manner.

Synthetic Aperture Radar (SAR), however, can penetrate clouds and acquire imagery day or night, which makes it particularly useful for flood disaster monitoring. NASS used the Copernicus Sentinel-1A and Sentinel-1B SAR for the first time to assess flooding impacts in Texas, Louisiana, and Florida during the Hurricane Harvey and Irma disaster events in 2017. With SAR, NASS was able to derive and provide a number of geospatial decision–support products to help the Federal Emergency Management Agency and others with immediate response and targeted recovery. These products included crop and pasture land inundated area maps and percentages of impacted crops, assessment reports, crop inundation raster layers, and wind swaths or surface winds overlaid onto crop areas identified from the NASS Cropland Data Layer (CDL) product. Access to and use of the satellite technology will allow NASS to provide rapid response to help with future extreme weather events.

Land Cover
- Cropland
- Inundated Cropland
- Inundated Pasture/Hay
- Other
- Pasture Hay
- Water

Photo: Flooded nursery in St. Lucie County, Florida. Credit: Monica Ozores-Hampton.

Photo: Flooded pasture in Polk County, Florida. Credit: Monica Ozores-Hampton.

Photo: Flooded Southwest Florida orange grove. Credit: Gene McAvoy on Twitter @SWFLVegMan.

Hurricane Irma results for Florida

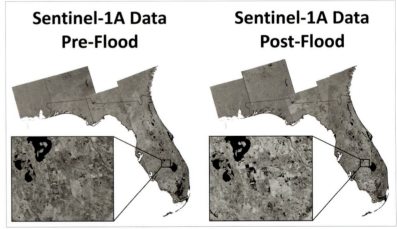

Sentinel-1A Data Pre-Flood

Sentinel-1A Data Post-Flood

Sentinel-1A data acquired over Florida before and after Hurrican Irma

CONTACT
Lee Ebinger
Lee.Ebinger@nass.usda.gov

SOFTWARE
ArcGIS Desktop 10.3

DATA SOURCES
Copernicus Sentinel-1A SAR imagery (July 31, 2017 to September 15, 2017), US Department of Agriculture/NASS 2016 Cropland Data Layer imagery and 2016 Cultivated Layer imagery,

National Oceanic and Atmosperic Admministration National Hurricane Center - hurricane track and wind swath radii

Florida photos by Monica Ozores-Hampton and Gene McAvoy; Texas photos by Joe Raedle/Getty Images, Staff Sargent Daniel J. Martinez of the US Air National Guard, and Dr. Josh McGinty of Texas A&M AgriLife Extension Service.

Land Cover

- **Cropland** (dark green)
- **Inundated Cropland** (orange-red)
- **Inundated Pasture/Hay** (yellow)
- **Other** (cream)
- **Pasture Hay** (brown)
- **Water** (blue)

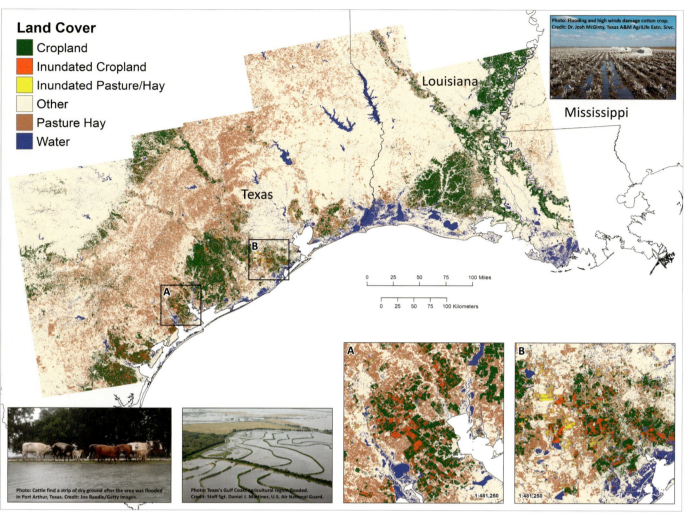

Louisiana

Mississippi

Texas

A

B

Photo: Flooding and high winds damage cotton crop. Credit: Dr. Josh McGinty, Texas A&M AgriLife Extn. Srvc.

0 25 50 75 100 Miles

0 25 50 75 100 Kilometers

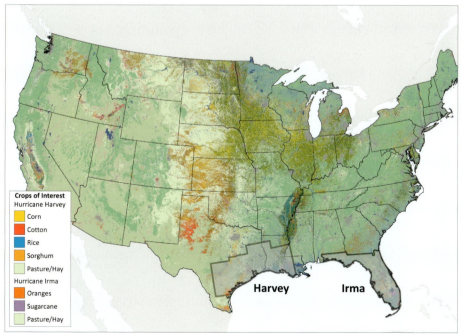

Photo: Cattle find a strip of dry ground after the area was flooded in Port Arthur, Texas. Credit: Joe Raedle/Getty Images.

Photo: Texas's Gulf Coast agricultural region flooded. Credit: Staff Sgt. Daniel J. Martinez, U.S. Air National Guard.

A

1:481,250

B

1:481,250

Hurricane Harvey results for portions of Louisana and Texas

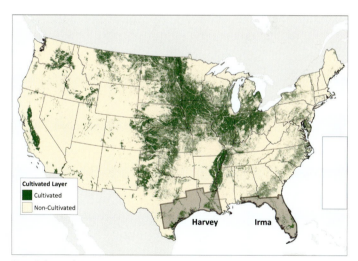

Crops of Interest

Hurricane Harvey
- Corn (yellow)
- Cotton (red)
- Rice (blue)
- Sorghum (orange)
- Pasture/Hay (light green)

Hurricane Irma
- Oranges (orange)
- Sugarcane (purple)
- Pasture/Hay (light green)

Harvey

Irma

2016 Cropland Data Layer

Cultivated Layer
- Cultivated (dark green)
- Non-Cultivated (cream)

Harvey

Irma

2016 Cultivated Layer

LAND SUITABILITY MAP FOR GROWTH OF HAZELNUTS IN SISIAN REGION, ARMENIA

Yerevan State University
Yerevan, Armenia
By Artak Piloyan

The purpose of this geospatial analysis was to find the most suitable areas for hazelnut growth in the Sisian region of Armenia. ArcGIS ModelBuilder was used for analysis procedures. Some of the initial criteria used to develop a suitability model included the distance from irrigation network and intercommunity roads, soil types, elevation, slope, aspect, the amount of annual precipitation, and the number of annual warm days and spring frost days.

CONTACT
Artak Piloyan
artakpiloyan@ysu.am

SOFTWARE
ArcGIS Desktop 10.2.1

DATA SOURCES
Center of Geodesy and Cartography, Syunik Marz Administration, Yerevan State University Chair of Cartography and Geomorphology, US Geological Survey

Legend

- Settlements
- Highways, Main roads
- Secondary roads
- Rivers
- Lakes

Land Suitability

- Unsuitable
- Mostly unsuitable
- Weakly suitable
- Suitable
- High suitable

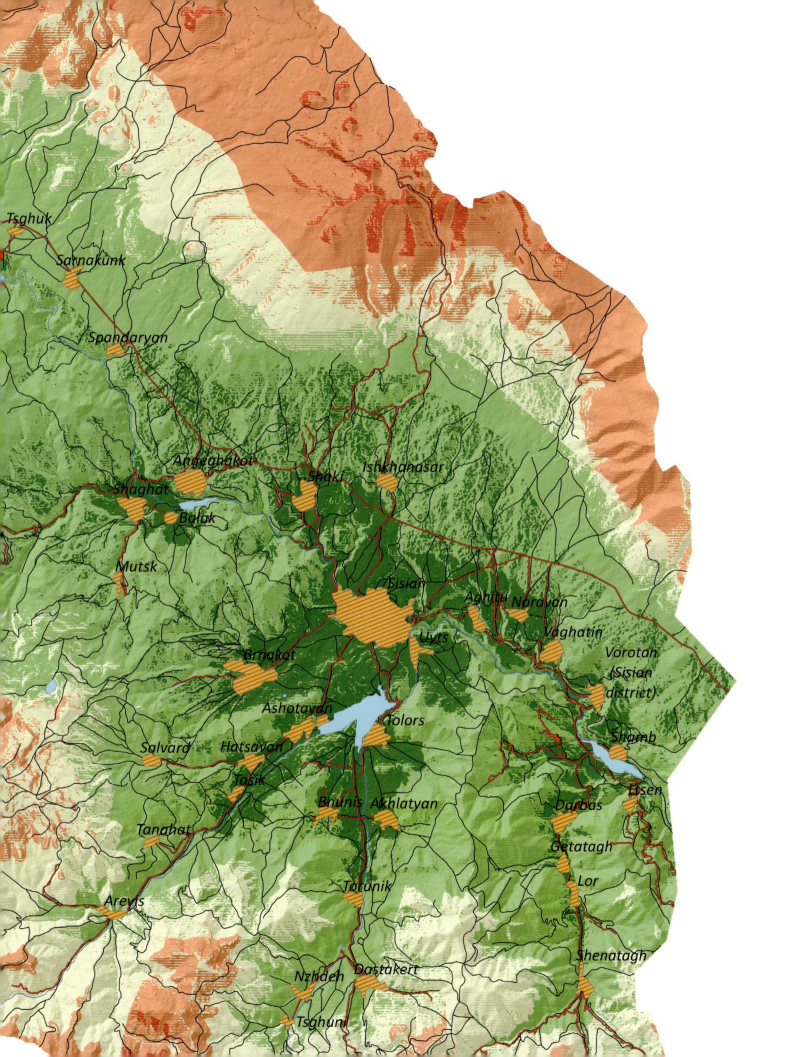

Tsghuk

Sarnakunk

Spandaryan

Angeghakot

Shaghat

Balak

Snaki

Ishkhanasar

Mutsk

Sisian

Aghiti Norayan

Vaghatin

Uyts

Vorotan
(Sisian
district)

Brnakot

Ashotavan

Tolors

Shamb

Salvard

Hatsavan

Tasik

Ltsen

Bnunis Akhlatyan

Darbas

Tanahat

Getatagh

Lor

Areyis

Torunik

Shenatagh

Nzhdeh

Dastakert

Tsghuni

RIPARIAN CATCHMENTS–
A FRAMEWORK
FOR AGRICULTURAL
CONSERVATION

USDA Agricultural Research Service
Amnes, Iowa, USA
By David James, Sarah Porter, Mark Tomer, and Jessy Van
Horn

The Agricultural Conservation Planning Framework (ACPF) proposes a conceptual framework, a high-resolution geospatial database, with ArcGIS Toolbox that can be used to achieve the potential broad-based benefits of precision agricultural conservation. The framework identifies options to locate multiple best management practices at the watershed, farm, and field levels. This framework also provides a planning resource to help watershed communities explore options to expand ecosystem services obtained from agricultural landscapes.

Riparian catchments on each side of the stream are analyzed independently using land use, soils, and terrain data. We hypothesize riparian catchments can better connect the potential downstream effects of landscape-based conservation actions.

CONTACT
David James
david.james@ars.usda.gov

SOFTWARE
ArcGIS Desktop 10.5.1

DATA SOURCES
US Department of Agriculture, Iowa State University GIS Facility, Iowa Department of Natural Resources

Courtesy of USDA, Iowa State University, and Iowa DNR.

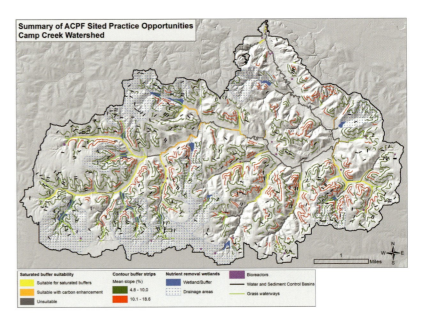

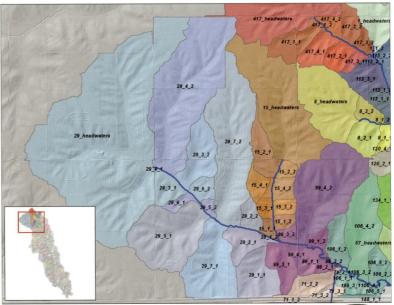

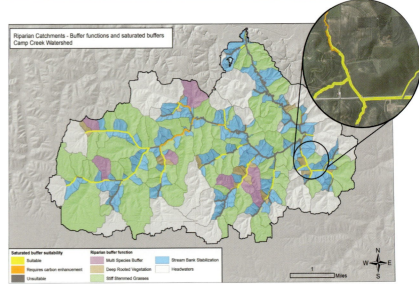

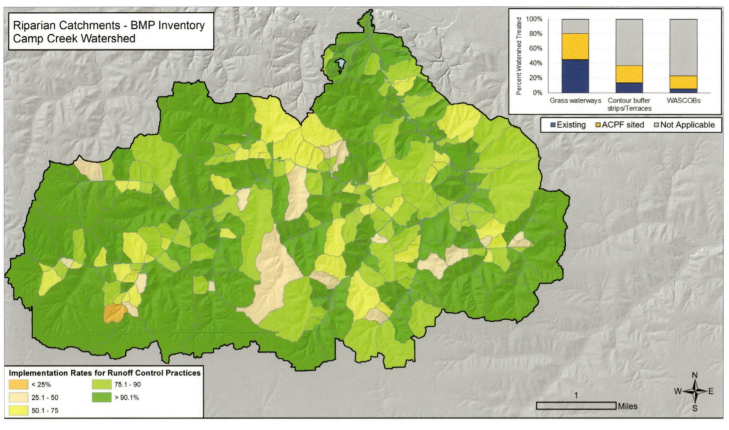

**Riparian Catchments - BMP Inventory
Camp Creek Watershed**

Percent Watershed Treated

Grass waterways | Contour buffer strips/Terraces | WASCOBs

■ Existing ☐ ACPF sited ☐ Not Applicable

Implementation Rates for Runoff Control Practices

- < 25%
- 25.1 - 50
- 50.1 - 75
- 75.1 - 90
- > 90.1%

1 Miles

N W E S

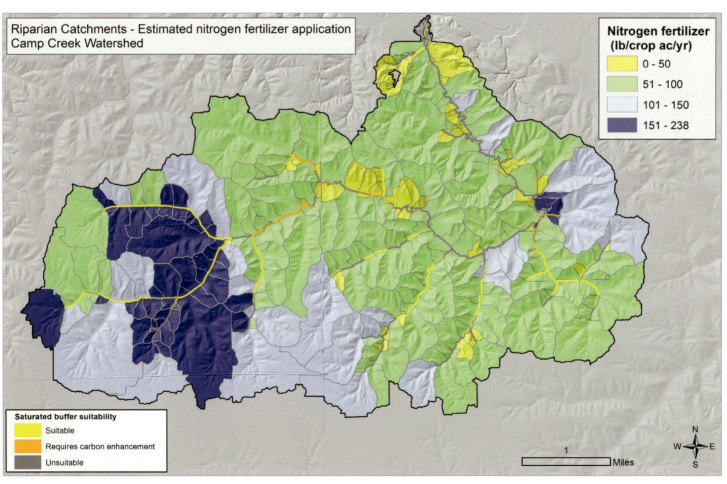

**Riparian Catchments - Estimated nitrogen fertilizer application
Camp Creek Watershed**

**Nitrogen fertilizer
(lb/crop ac/yr)**

- 0 - 50
- 51 - 100
- 101 - 150
- 151 - 238

Saturated buffer suitability

- Suitable
- Requires carbon enhancement
- Unsuitable

1 Miles

N W E S

2017 WILDFIRE SEASON: A RECORD-SETTING FIRE SEASON IN THE PROVINCE OF BC

GeoBC
Victoria, British Columbia, Canada
By Sarah MacGregor, Natalie Work Advisors: Nancy Liesch, Susan Westmacott, Steve Hann, Stephen Sutherland, and Ben Arril

The summer of 2017 will be remembered as one of the worst wildfire seasons in British Columbia's history. The provincial declaration of emergency was the longest in provincial history, lasting seventy days from July 7 to September 15. At peak activity, over 4,700 personnel were engaged in fighting wildfires across British Columbia, including 2,000 contract personnel from the forest industry and over 1,200 personnel from outside the province. The scale of the 2017 fire season was larger than anything on British Columbia's record books, with over 1 million hectares burned.

At the request of an assistant deputy minister, the 2017 Wildfire Season map was created to commemorate and acknowledge the impact of the fire season across the province and to honor the great efforts of the people who responded to the emergency.

This map summarizes the extent of the fires in the province with respect to the number of hectares burned, the cost, the number of fires, the number of displaced people, and other summary statistics of this historical season. Multiple copies were printed for display in the provincial legislature, executive offices, and wildfire centers across the province.

CONTACT
Sarah MacGregor
Sarah.MacGregor@gov.bc.ca

SOFTWARE
ArcGIS Desktop 10.3, Adobe Photoshop CC 2017

DATA SOURCES
GeoBC Evacuation Orders and Alerts, BC Wildfire Service Fire Perimeters and Fire Points, Government of Canada: 1:1 million base (cartographically enhanced by GeoBC)

Courtesy of GeoBC.

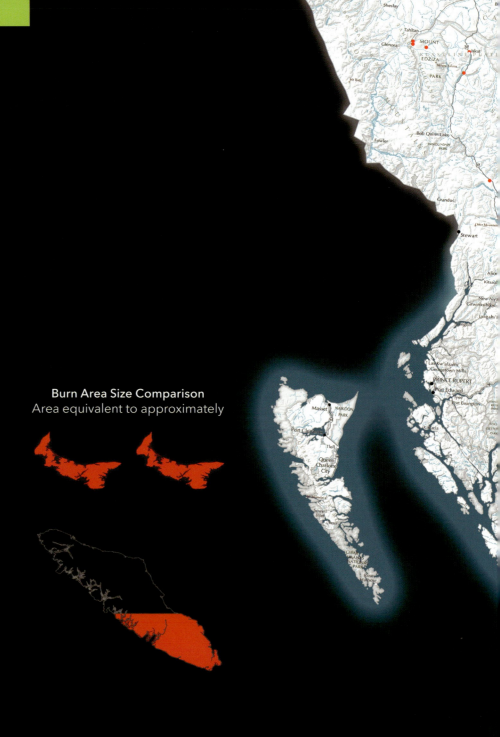

Burn Area Size Comparison
Area equivalent to approximately

VULNERABILITY OF UPLAND FOREST TO CHANGING CLIMATE

City of Seattle
North Bend, Washington, USA
By Mark Joselyn and Rolf Gersonde

Changing climate presents challenges to land managers engaged in active restoration activities. This is certainly true for the scientists working to implement the 50-year Cedar River Watershed Habitat Conservation Plan (HCP) for Seattle Public Utilities. A landscape analysis of how vulnerable upland forests are to the impacts of climate change was developed to guide restoration efforts. This spatially explicit model of vulnerability is intended to guide efforts and reduce the uncertainty in restoration outcomes.

To address these issues, we followed a model of vulnerability assessment that leans on the National Wildlife Federation's vulnerability assessment guide. In this framework, exposure and sensitivity (the anticipated response of existing flora) of upland forest ecosystems along the western slopes of the cascades are the drivers of vulnerability. The response of an ecosystem ranges from resilient, less stressed systems to those that will experience higher and increased stress.

CONTACT
Mark Joselyn
mark.joselyn@seattle.gov

SOFTWARE
ArcGIS Desktop 10.3

DATA SOURCES
Lidar

Courtesy of City of Seattle.

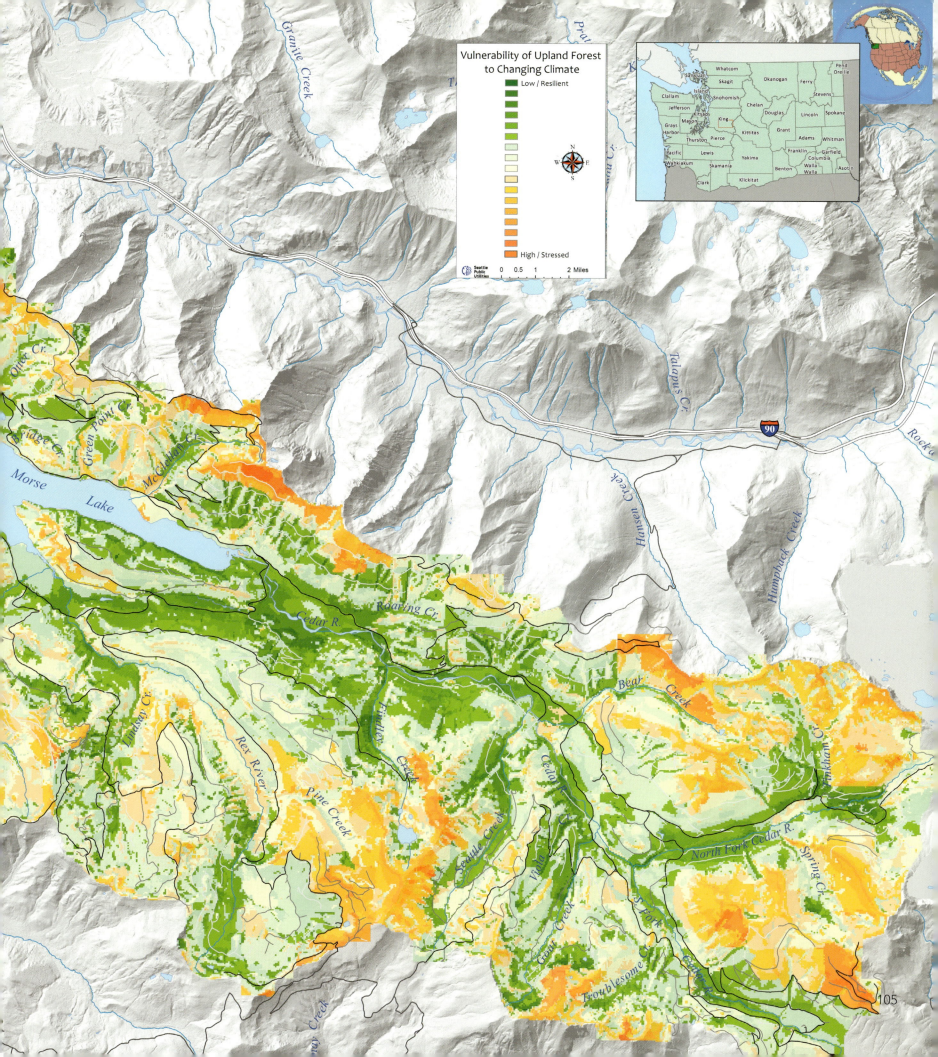

Vulnerability of Upland Forest to Changing Climate

Low / Resilient

High / Stressed

N
W E
S

0   0.5   1   2 Miles

Seattle
Public
Utilities

FOREST BLOCK ANALYSIS AND PRIORITIZATION IN ALBEMARLE COUNTY, VA

County of Albemarle, Virginia
Charlottesville, Virginia, USA
By Andrew Walker

Albemarle County, Virginia, recently completed an analysis of forest habitat to support its Biodiversity Action Plan. This map summarizes, at a high level, the prioritization of large forest blocks based on multiple metrics that estimate their conservation value.

Large, contiguous patches of forest (the predominant type of natural habitat in Albemarle County) serve as a critical component to protecting biodiversity and ecological resiliency. In Albemarle County, there is a particularly strong legacy of preserving water quality and quantity by protecting natural and rural areas.

In this map, forest blocks are categorized into two main tiers, and Conservation Focus Areas were identified by analyzing the areas where large forest blocks with the greatest conservation value are clustered. Three areas of the county stand out as having significant conservation value. These Conservation Focus Areas will be a focus of conservation efforts during the next five years.

CONTACT
Andrew Walker
r.andrewwalker@gmail.com

SOFTWARE
ArcGIS Desktop 10.3

DATA SOURCES
Albemarle County; Green Infrastructure Center; US Fish and Wildlife Service (NWI)

Courtesy of County of Albemarle, Virginia.

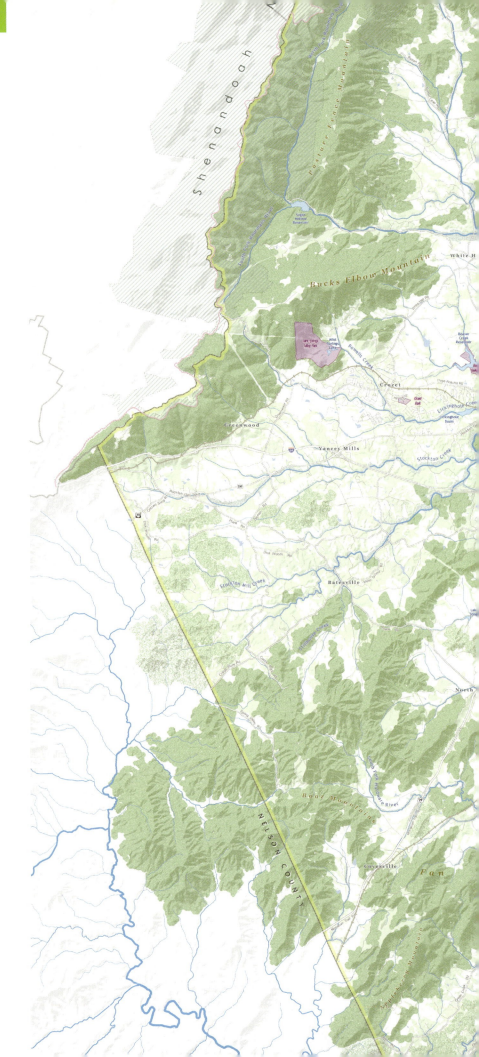

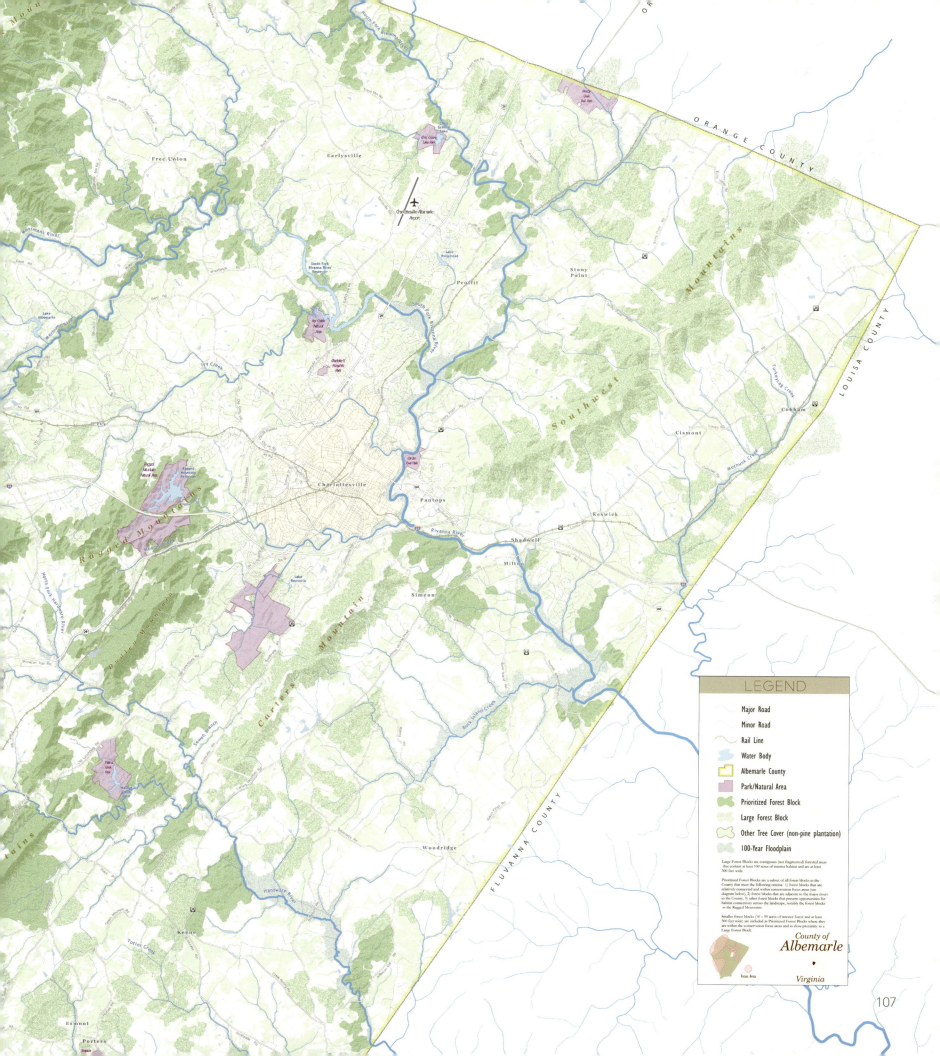

UTAH TRUST LANDS AND THE REDUCTION OF BEARS EARS NATIONAL MONUMENT

Trust Lands Administration
Salt Lake City, Utah, USA
By Kate Staley

The School and Institutional Trust Lands Administration (SITLA) is an independent state agency of Utah. The agency was created in 1994 to manage parcels of land granted to the state by Congress in 1894. Today, SITLA administers approximately 4.5 million mineral acres and approximately 3.4 million surface acres of trust lands to provide funding and support to the permanent school fund.

This map was created to show trust lands parcels impacted by the designation and amendment of Bears Ears National Monument in southeastern Utah. The Bears Ears National Monument was designated on December 28, 2016, by President Barack Obama to protect public lands. The original boundary included approximately 1.35 million acres (approximately 116,000 trust lands acres). On December 4, 2017, the Bears Ears National Monument was amended by President Donald J. Trump to shrink the boundary to approximately 202,000 acres (approximately 26,000 trust lands acres).

CONTACT
Katherine Staley
katestaley@utah.gov

SOFTWARE
ArcGIS Desktop 10.4.1

DATA SOURCES
Utah Automated Geographic Reference Center, Bureau of Land Management, US Geoglogical Survey

Courtesy of State of Utah School and Institutional Trust Lands Administration.

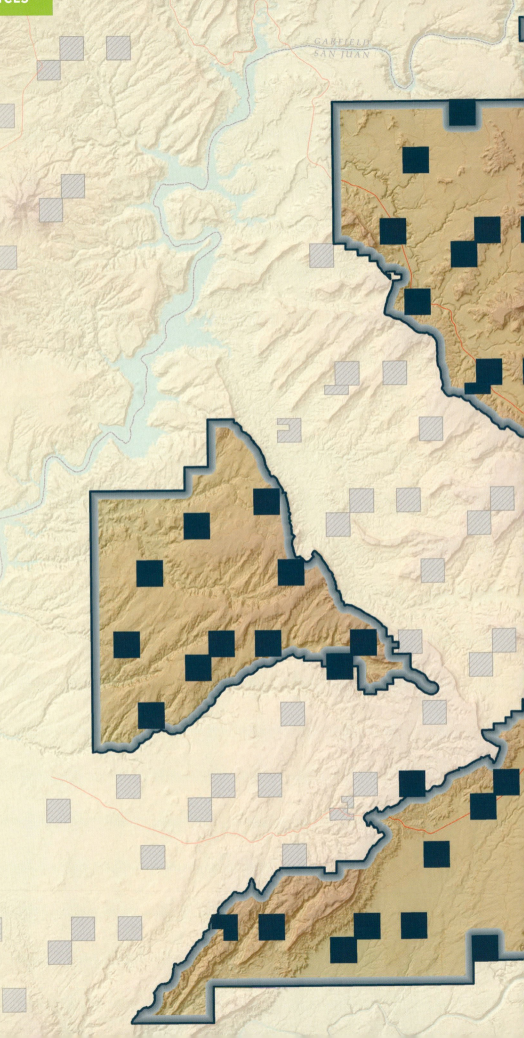

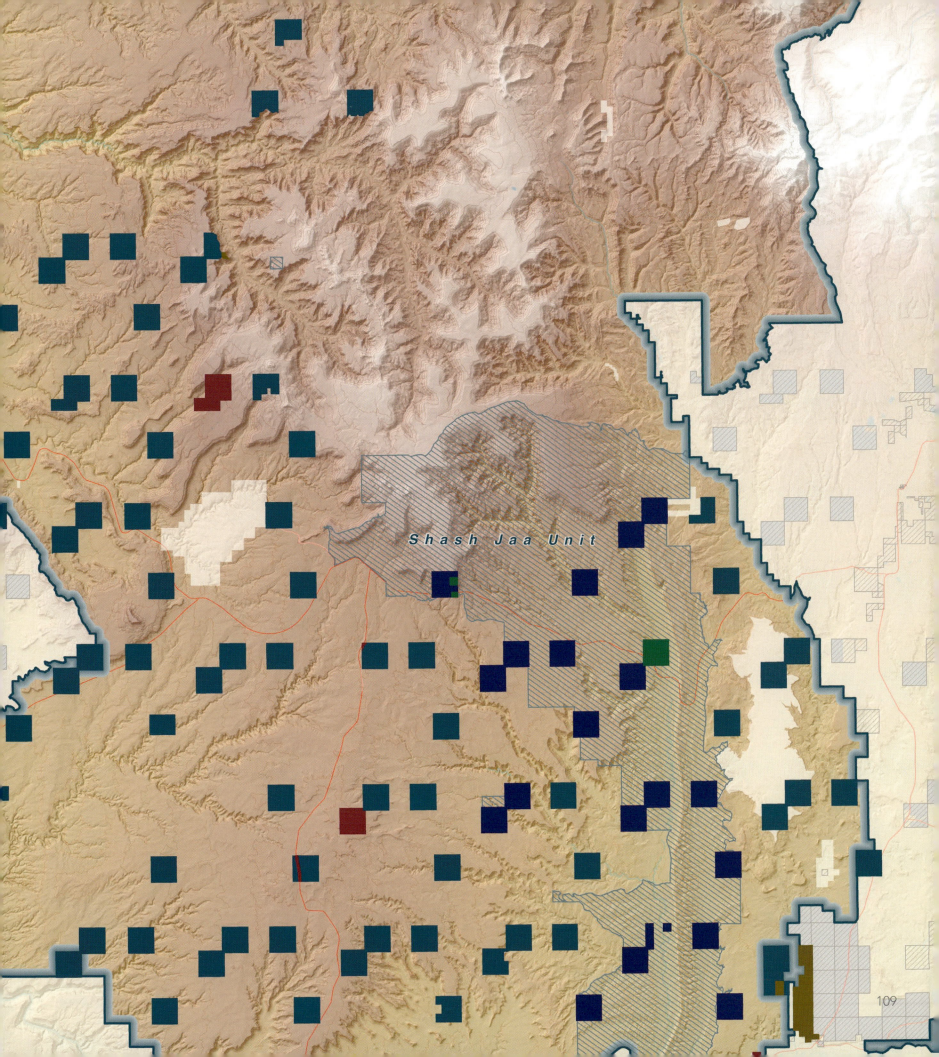

Shash Jaa Unit

THE OPENING OF THE CENTRAL ATLANTIC

Getech Group, PLC
Leeds, West Yorkshire, United Kingdom
By Robert Bailiff, David Blackledge, and Kevin Mckenna

This maps series shows the evolution of the Central Atlantic region between the Late Triassic and the Middle Jurassic from a tectonic and topographic perspective. In the Triassic, rifting begins between Africa and North America, which are both part of the supercontinent of Pangaea at this time. As rifting progresses, the central Atlantic magmatic province begins to form, with volcanism affecting vast areas of North America, Africa, and South America and large-scale eruptions of lavas and dikes occurring for 2 to 3 million years. Thinning of the crust causes the region to subside, resulting in both the formation of a broad sag zone between Africa and North America and the invasion of seawater from the north into the region. As rifting continues, the zone of thinned crust widens, until the two continents gradually split apart in the Middle Jurassic, and the Central Atlantic Ocean forms between them.

CONTACT
Kevin Mckenna
kmk@getech.com

SOFTWARE
ArcGIS Desktop 10.3

DATA SOURCES
Getech Group plc

Courtesy of Getech Group plc.

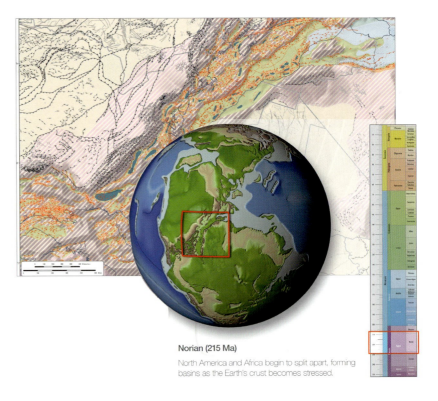

Norian (215 Ma)
North America and Africa begin to split apart, forming basins as the Earth's crust becomes stressed.

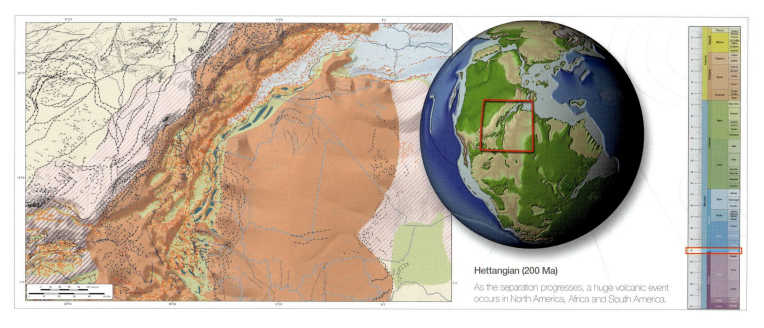

Hettangian (200 Ma)

As the separation progresses, a huge volcanic event occurs in North America, Africa and South America.

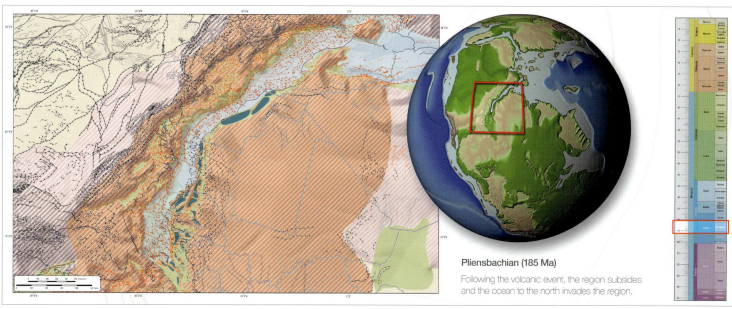

Pliensbachian (185 Ma)

Following the volcanic event, the region subsides and the ocean to the north invades the region.

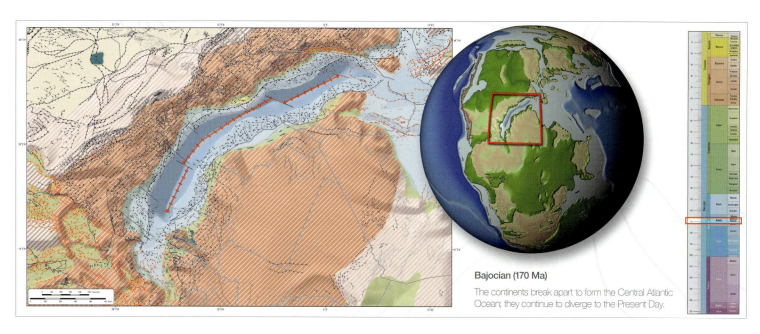

Bajocian (170 Ma)

The continents break apart to form the Central Atlantic Ocean; they continue to diverge to the Present Day.

UTAH EARTHQUAKES (1850 TO 2016) AND QUATERNARY FAULTS

Utah Department of Natural Resources
Salt Lake City, Utah, USA
By Steve D. Bowman and Walter J. Arabasz; cartography by Gordon E. Douglass and Jay C. Hill

This map shows earthquakes known to have occurred within and surrounding Utah from 1850 through December 2016 and mapped Quaternary faults considered to be earthquake sources. The faults shown on the map have been sources of large earthquakes (about magnitude 6.5 or greater) during the Quaternary Period (past 2.6 million years) and are the most likely sources of large earthquakes in the future. Most small-to-moderate size earthquakes plotted on the map are "background" earthquakes not readily associated with known faults and of a size generally below the threshold of surface faulting (about magnitude 6.5). Buried or unmapped secondary faults are likely sources of much of the background seismicity.

CONTACT
Gordon Douglass
gdouglass@utah.gov

SOFTWARE
ArcGIS Desktop 10.6

DATA SOURCES
Utah Geological Survey, University of Utah

Courtesy of Utah Department of Natural Resources.

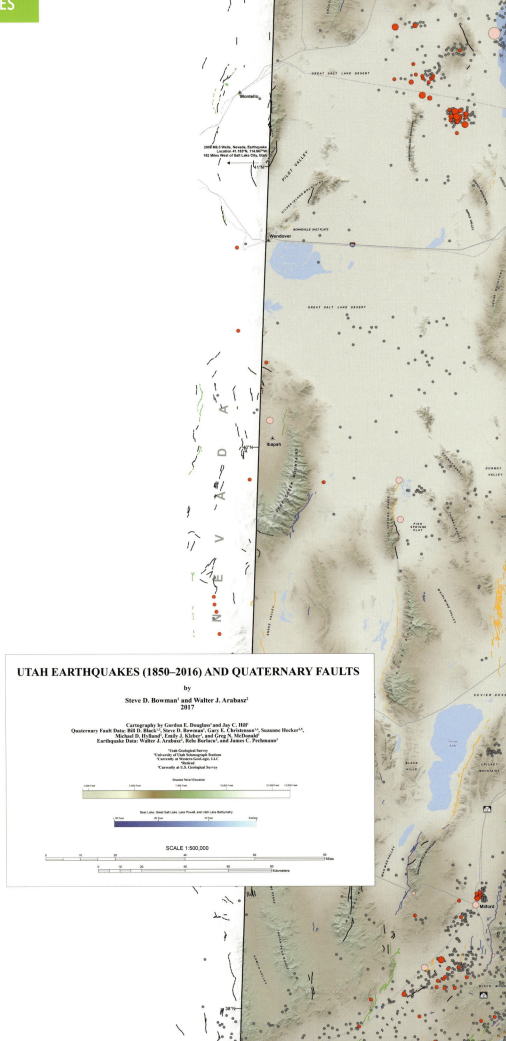

UTAH EARTHQUAKES (1850–2016) AND QUATERNARY FAULTS

by

Steve D. Bowman[1] and Walter J. Arabasz[2]
2017

Cartography by Gordon E. Douglass[1] and Jay C. Hill[1]
Quaternary Fault Data: Bill D. Black[1,3], Steve D. Bowman[1], Gary E. Christenson[1,4], Suzanne Hecker[1,5],
Michael D. Hylland[1], Emily J. Kleber[1], and Greg N. McDonald[1]
Earthquake Data: Walter J. Arabasz[2], Relu Burlacu[2], and James C. Pechmann[2]

[1]Utah Geological Survey
[2]University of Utah Seismograph Stations
[3]Currently at Western GeoLogic, LLC
[4]Retired
[5]Currently at U.S. Geological Survey

SCALE 1:500,000

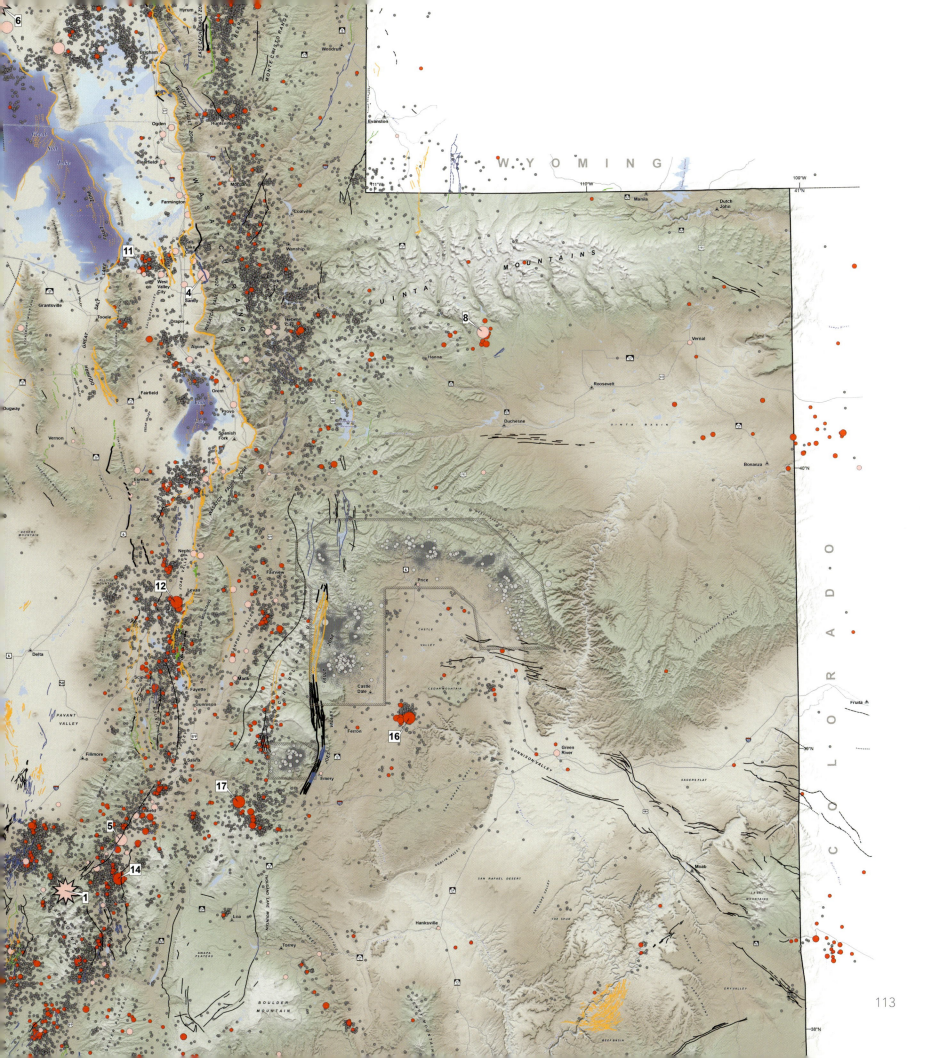

GROUNDWATER MODEL DEVELOPMENT AND GRAPHICAL PRESENTATION WITH ARCGIS

Ramboll
Emeryville, California, USA
By Alka Singhal and Lisa A. Taylor

Groundwater flow and contaminant fate and transport modeling are important components of most aquifer remediation studies. Development of a groundwater model involves creating a 3D model grid representing the model domain, setting the hydrogeological properties of the grid cells, and assigning boundary and initial conditions to control the flow of groundwater through the model.

In the environmental industry, models are used to calculate groundwater flow trajectories and to evaluate the extent of capture of groundwater pumping systems. The outputs of the model can be post-processed to be viewed in ArcGIS, allowing modeling outcomes to be evaluated in the context of other spatial datasets, such as property boundaries and contaminant plume extents.

CONTACT
Alka Singhal
asinghal@ramboll.com

SOFTWARE
ArcGIS Spatial Analyst, Leapfrog® Hydro

DATA SOURCES
Esri, US Geological Survey, Clark County, Ramboll

Copyright © Ramboll.

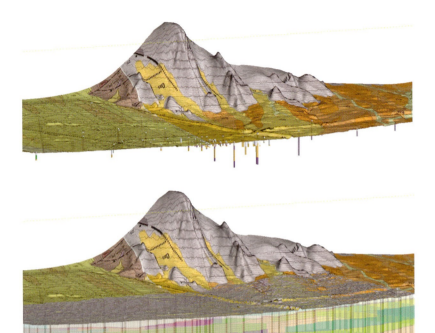

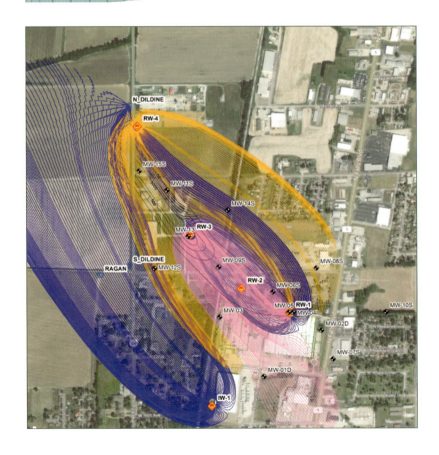

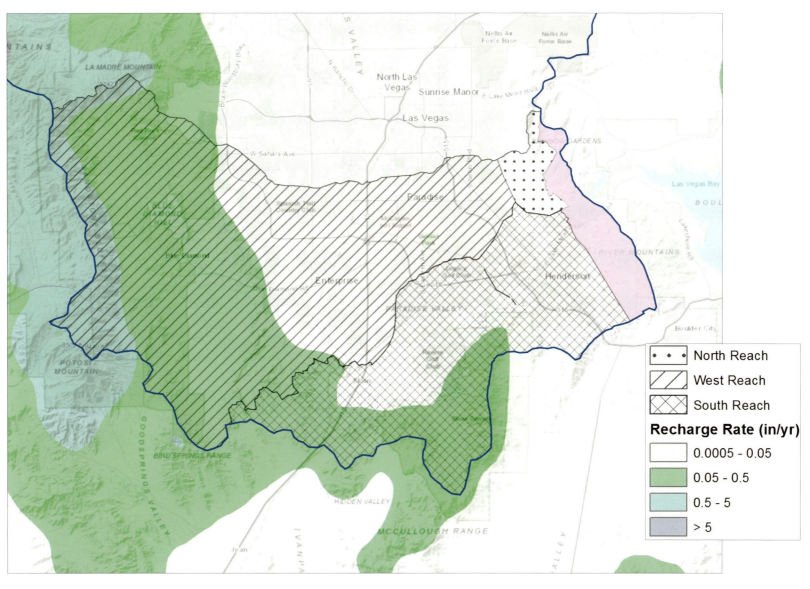

Recharge Rate (in/yr)

- North Reach
- West Reach
- South Reach

- 0.0005 - 0.05
- 0.05 - 0.5
- 0.5 - 5
- > 5

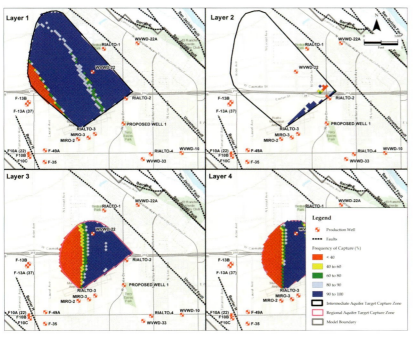

Legend

- Production Well
- Faults

Frequency of Capture (%)
- < 40
- 40 to 60
- 60 to 80
- 80 to 90
- 90 to 100
- Intermediate Aquifer Target Capture Zone
- Regional Aquifer Target Capture Zone
- Model Boundary

STATEWIDE (1 FT.) IMPERVIOUS SURFACE TO SUPPORT STORMWATER MAPPING

University of Connecticut
Storrs, Connecticut, USA
By Emily Wilson, Cary Chadwick, and David Dickson

The MS4 (municipal separate storm sewer system) general permit is a regulation issued by the state of Connecticut that applies to 121 of Connecticut's 169 municipalities. In Connecticut, towns are required to focus their efforts on three priority areas, one of which is areas of high impervious cover (basins greater than 11 percent directly connected impervious area). Percent impervious area within each small watershed identifies the areas of town that should be the focus of stormwater management activities. The impervious surface area was determined from statewide, one-foot impervious surface data. The thematic raster layer includes three classes: building, roads, and other impervious and is available on Connecticut Environmental Conditions Online (CT ECO), a website that makes Connecticut's geospatial information available.

CONTACT
Emily Wilson
emily.wilson@uconn.edu

SOFTWARE
ArcGIS Pro 2.1.2, ArcGIS Desktop 10.5, Microsoft PowerPoint

DATA SOURCES
University of Connecticut Center for Land Use Education and Research, CT ECO, Connecticut Department of Energy and Environmental Protection

Courtesy of University of Connecticut.

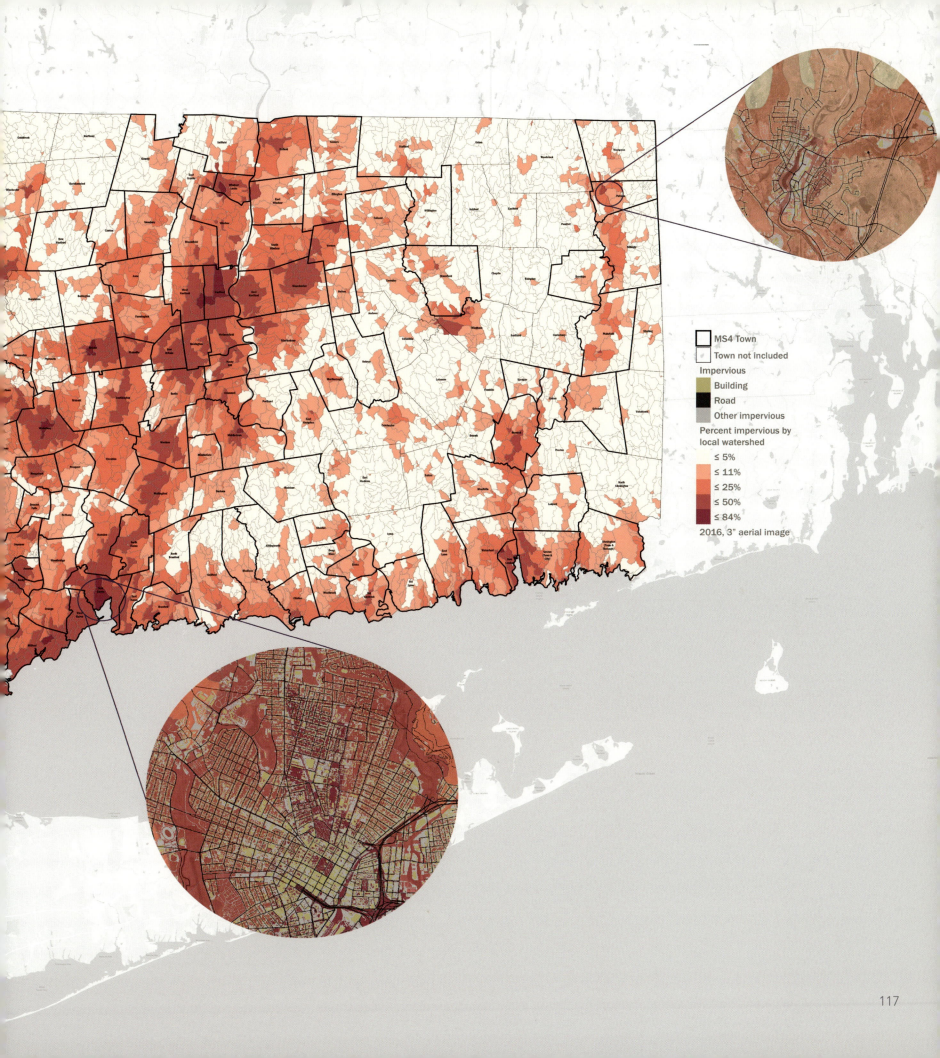

MS4 Town

Town not included

Impervious

Building

Road

Other impervious

Percent impervious by
local watershed

≤ 5%

≤ 11%

≤ 25%

≤ 50%

≤ 84%

2016, 3" aerial image

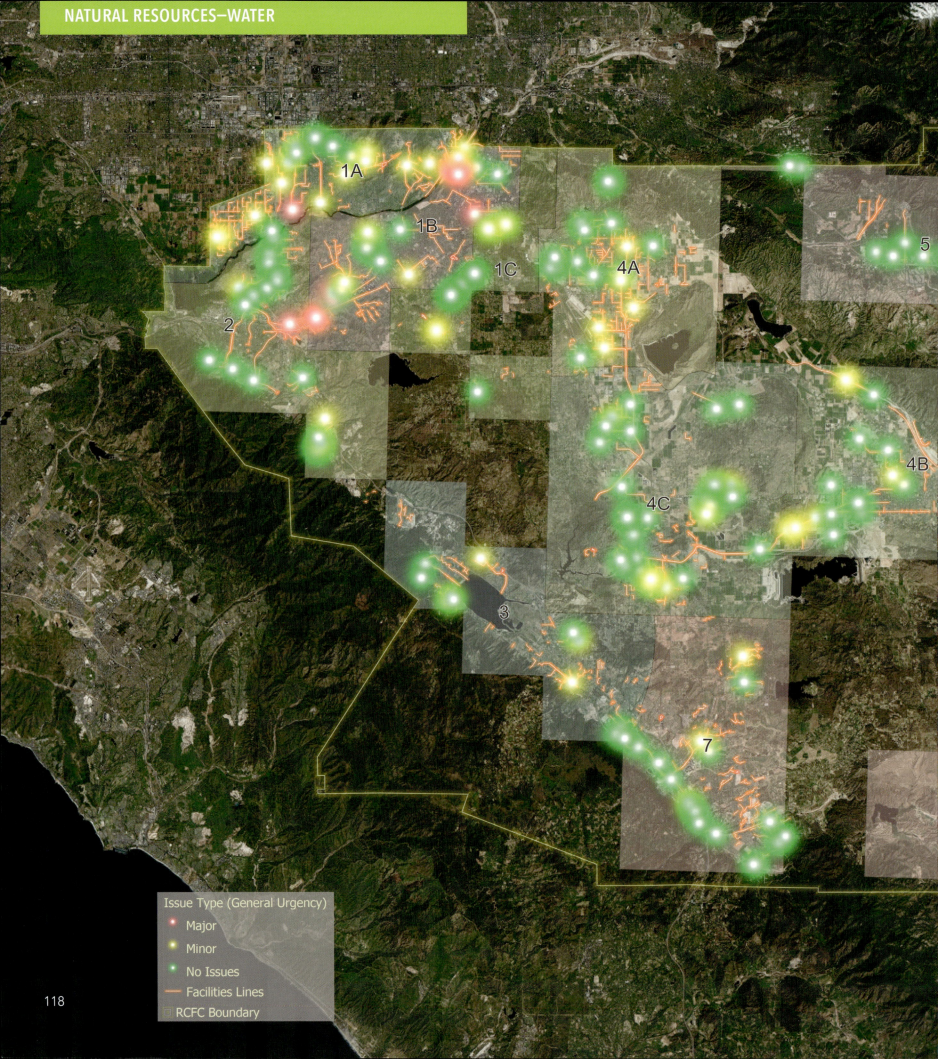

1A

1B

1C

2

3

4A

4B

4C

5

7

Issue Type (General Urgency)
- Major
- Minor
- No Issues
- Facilities Lines
- RCFC Boundary

STORM PATROL INSPECTION

Riverside County Flood Control and Water Conservation
District
Riverside, California, USA
By Aldous Tsang

During a severe rain event, Riverside County Flood Control
and Water Conservation District (RCFCWCD) storm patrol
teams are sent out to inspect facilities' conditions. The
map shows the general status of the facilities (e.g., basins,
dams, and channels). Using Esri's Survey123 application,
a question on the survey form asks about the general
issue of the facility with the following choices of "Major,"
"Minor," or "No Issue." Once the survey is submitted, the
data is mapped to ArcGIS Online. Management back at
RCFCWCD is then able to see where maintenance crew
should be deployed based on the urgency in real-time.
The ability to make real-time decisions helped RCFCWCD
allocate proper resources to the most urgent areas, thus
protecting the public from potential risks.

CONTACT
Aldous Tsang
atsang@RIVCO.ORG

SOFTWARE
ArcGIS Pro 2.2

DATA SOURCES
RCFCWCD

*Courtesy of Riverside County Flood Control and Water
Conservation District.*

STORM SURGE IMPACT ON GNHWPCA FACILITIES

GNHWPCA
New Haven, Connecticut, USA
By Ricardo Ceballos, PE

The Greater New Haven Water Pollution Control Authority (GNHWPCA) is a regional sewer authority in Connecticut that serves the city of New Haven and the towns of East Haven, Hamden, and Woodbridge. GNHWPCA oversees the operation and maintenance of an extensive sewer system that includes 555 miles of sewer pipes, 30 pumping stations, and an advanced secondary wastewater treatment plant with a capacity of 40 million gallons per day.

Storm surge is "often the greatest threat to life and property from a hurricane," according to the National Hurricane Center. The greatest impact of any storm surge will occur during mean high tide periods that cause extreme flooding elevations. The combination of storm surge and the astronomical tide is known as storm tide. The impact of the tide on any predicted storm surge is significant because it adds in our area up to 6.15 feet of water to the storm surge.

GNHWPCA developed the storm surge impact map to show the analysis results of its service area's potential vulnerability to the storm tide under different storm surge predictions. The map shows all the pump stations and wastewater treatment facilities that are impacted during any storm surge and mean high tide event. The map also shows for reference the recorded high-water elevations of previous hurricanes in our area.

CONTACT
Ricardo Ceballos
rceballos@gnhwpca.com

SOFTWARE
ArcGIS Desktop 10.4

DATA SOURCES
GNHWPCA, National Oceanic and Atmospheric Administration, National Hurricane Center

Courtesy of GNHWPCA.

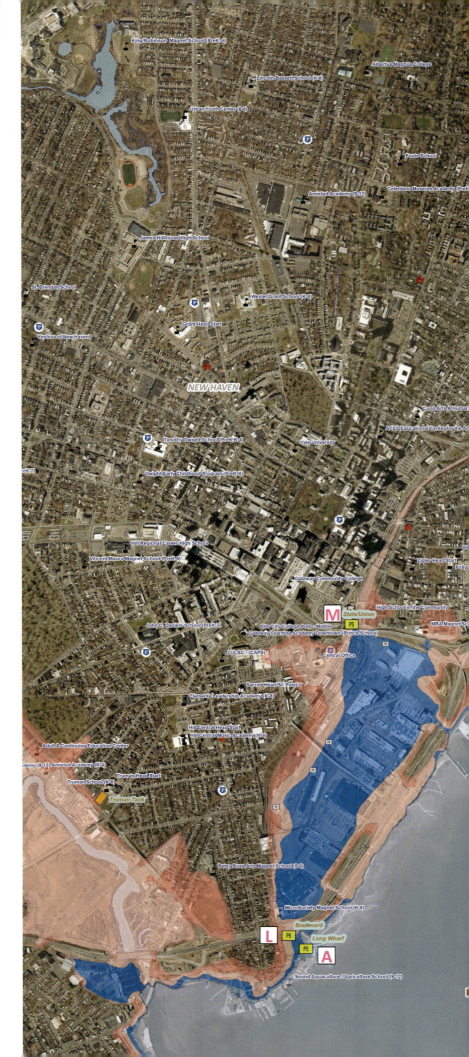

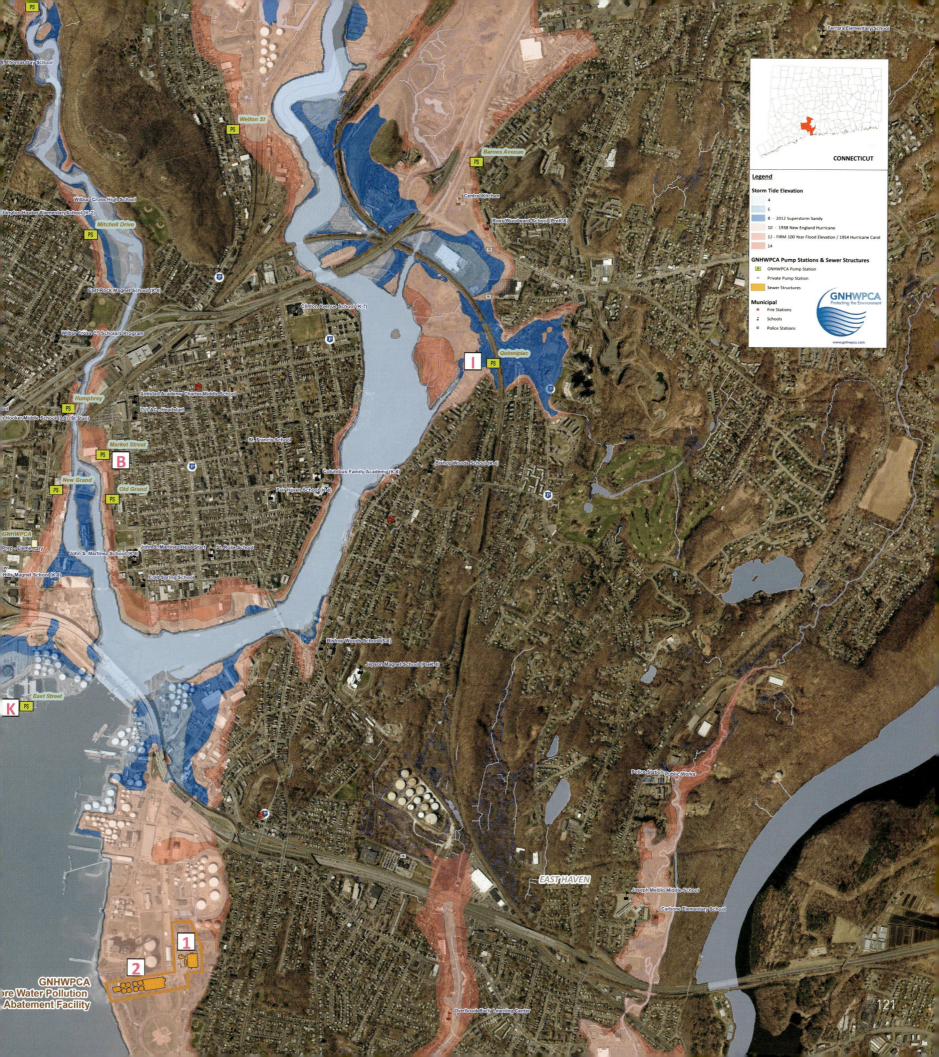

THE SACRAMENTO/SAN JOAQUIN DELTA

US Bureau of Reclamation
Sacramento, California, USA
By Tom Heinzer, Diane Williams, and David Mooney

The map of the Sacramento-San Joaquin Delta shows the complex network of waterways, islands, and facilities that comprise the hub of California's Federal Central Valley Project (CVP) and State Water Project serving 30 million people and irrigating millions of acres of agricultural land. As operator of the CVP, the Bureau of Reclamation is a senior partner with other federal, state, and local agencies, water and power stakeholders, commercial and sport fishing groups, the environmental community, and the general public in effectively meeting the multiple conflicting demands on water and ecosystem management.

The relationships between water supply in the north, pumps in the south, facilities-to-route flows, and the various monitoring and regulatory compliance locations are key to understanding better ways to maximize scarce water supplies and protect natural resources.

CONTACT
Thomas Heinzer
theinzer@usbr.gov

SOFTWARE
ArcGIS Suite

DATA SOURCES
Delta Legal Boundary, US Geological Survey, US Bureau of Reclamation, Esri

Courtesy of US Bureau of Reclamation.

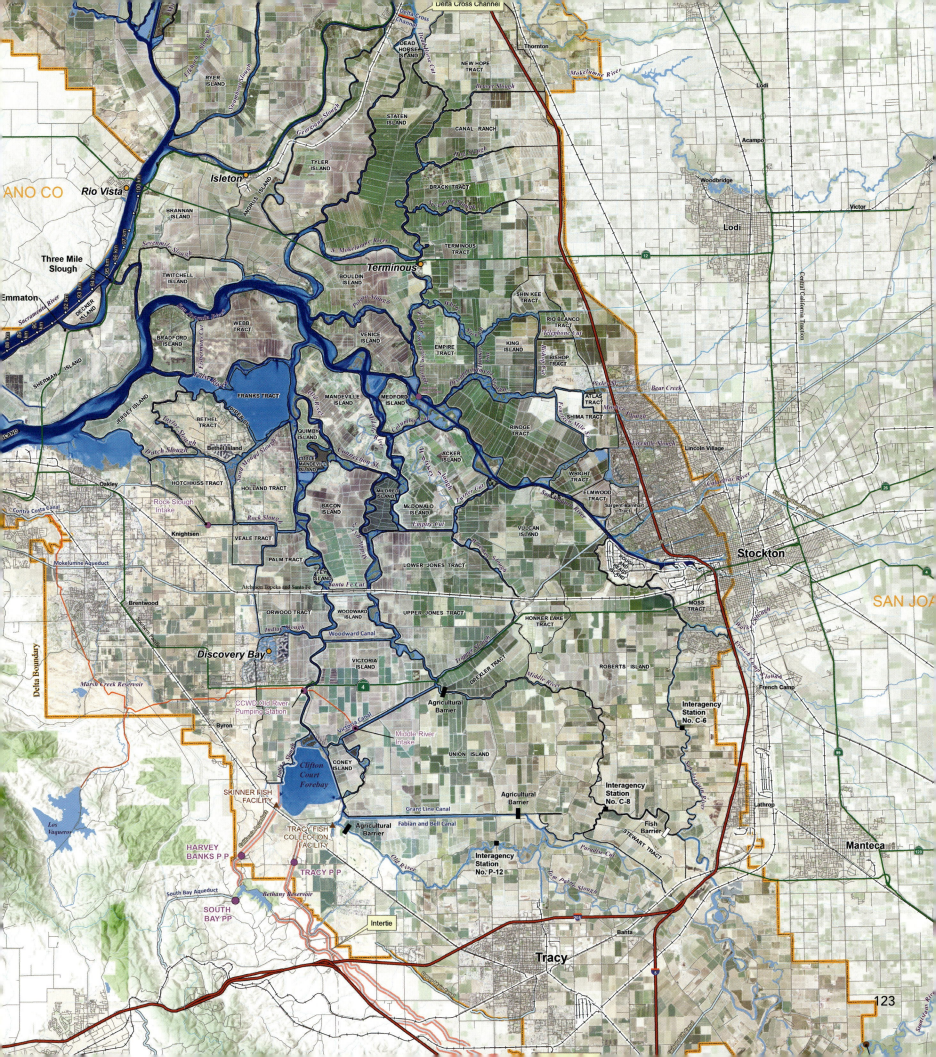

123

OREGON MPO TRANSPORTATION MOBILITY BROCHURES

Oregon Metro
Portland, Oregon, USA
By Julie Stringham and Matthew Hampton

In Oregon, metropolitan areas and their respective planning organizations (MPOs) are defined as discrete jurisdictions, but their increasingly interdependent economies, land use, and growing populations bring about the need to look beyond jurisdictional boundaries when addressing transportation and issues related to land use.

With the goal of discovering and better understanding these issues, a series of informational pamphlets were created for the Oregon Metropolitan Planning Organization Consortium (OMPOC) to highlight each MPO's unique geographic situation.

To achieve this, the pamphlets contain a series of maps which display a wide variety of data. Land cover, population, and transportation maps are used to situate MPOs within their greater regions. Daily traffic volumes and forty-five-minute travelsheds demonstrate the general flow of traffic and major access routes to the MPO from the surrounding region, while population and land cover maps illustrate population density and its effect on land cover and the configuration of developed land within MPO boundaries.

CONTACT

Matthew Hampton
matthew.hampton@oregonmetro.gov

SOFTWARE

ArcGIS Desktop, Adobe Illustrator, Adobe InDesign, Adobe Photoshop

DATA SOURCES

Metro Regional Land Information System, US Census Bureau, US Geological Survey Gap Report

Courtesy of Oregon Metro.

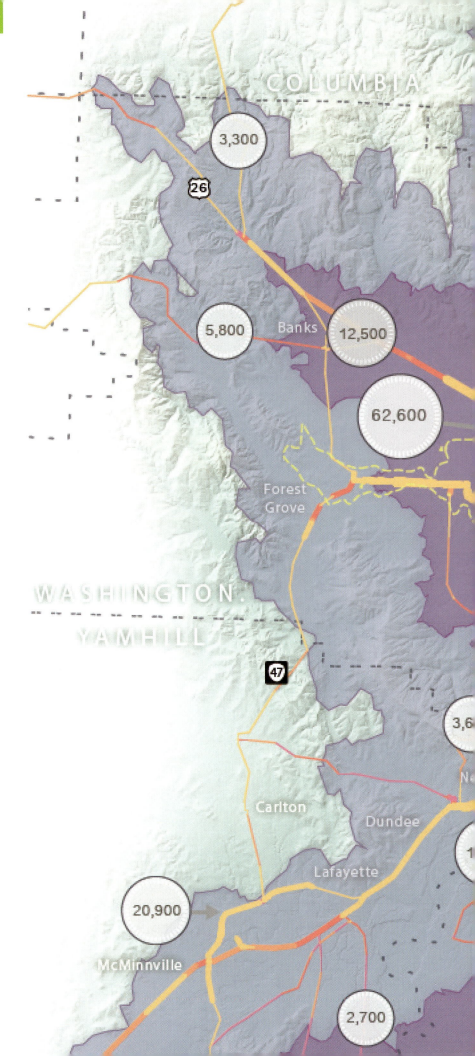

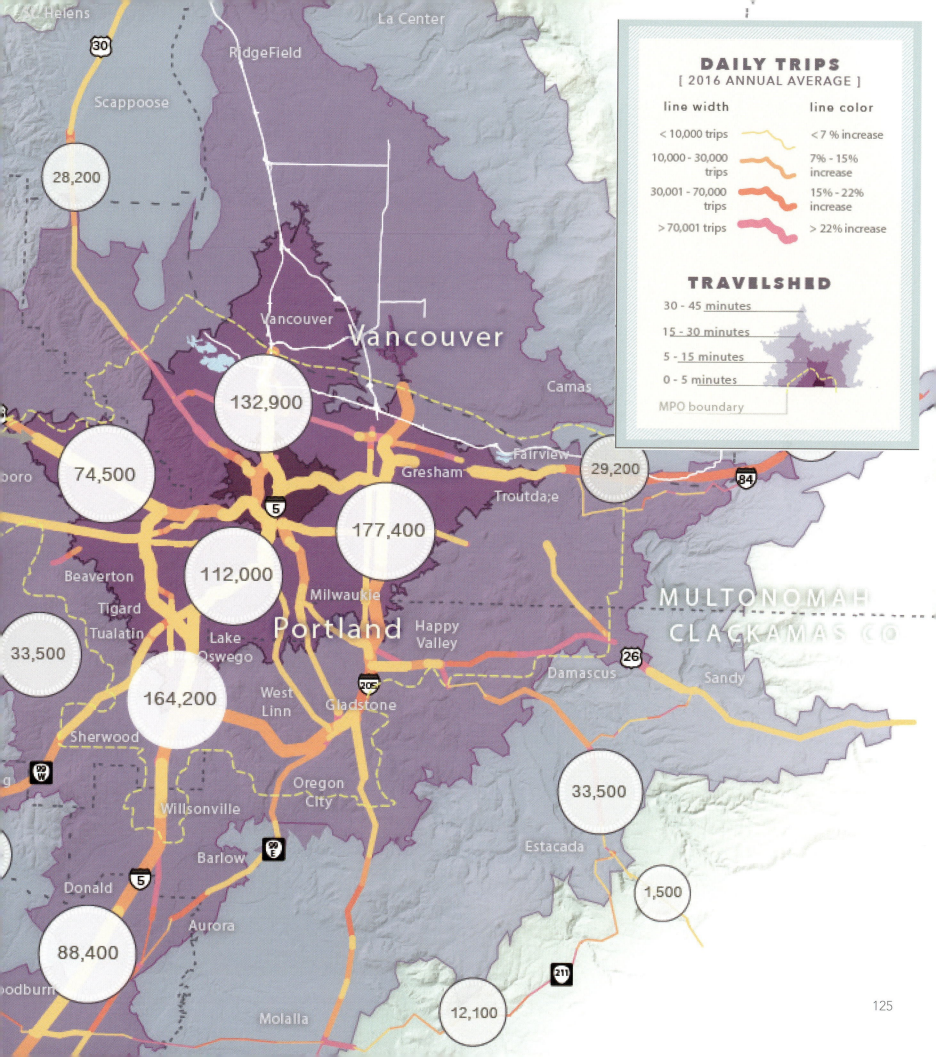

DAILY TRIPS
[2016 ANNUAL AVERAGE]

line width
- < 10,000 trips
- 10,000 - 30,000 trips
- 30,001 - 70,000 trips
- >70,001 trips

line color
- <7 % increase
- 7% - 15% increase
- 15% - 22% increase
- > 22% increase

TRAVELSHED
- 30 - 45 minutes
- 15 - 30 minutes
- 5 - 15 minutes
- 0 - 5 minutes
- MPO boundary

St. Helens
La Center
RidgeField
Scappoose
Vancouver
Vancouver
Camas
Fairview
Gresham
Troutda;e
Beaverton
Tigard
Tualatin
Milwaukie
Portland
Happy Valley
Damascus
Sandy
Lake Oswego
West Linn
Gladstone
Sherwood
Oregon City
Willsonville
Estacada
Barlow
Donald
Aurora
Molalla

MULTONOMAH
CLACKAMAS CO

28,200
132,900
29,200
74,500
177,400
112,000
33,500
164,200
33,500
1,500
88,400
12,100

UTILIZING ESRI APPLICATIONS TO ENHANCE DATA COLLECTION FOR BUILDING INSPECTIONS

Langan
Philadelphia, Pennsylvania, USA
By Michael Hagan

Esri's Collector and Survey123 were instrumental in maximizing efficiency in data collection at an active chemical manufacturing facility. Collector was used to conduct building inspections and document areas of concern for possible vapor intrusion pathways due to soil and groundwater contamination at the site. Survey123 was also used in conjunction with Collector to inventory the location of chemicals stored at the facility, how they are stored, and the overall condition. Due to the potential for exposure to harmful vapors, sub-slab soil gas, indoor air, and ambient air samples were taken to determine if a vapor mitigation system should be installed.

CONTACT
Michael Hagan
mhagan@langan.com

SOFTWARE
ArcGIS Pro, ArcGIS Online, Collector, Survey123

DATA SOURCES
Site inspection performed and data collected by Langan, May 2018; all other layers based on survey completed by Langan, October 2007

Courtesy of Langan.

CHEMICAL MANUFACTURING FACILITY

STREAM

Legend

Vapor Intrusion Mitigation Observation Points

Chemical Storage Points

Crack Not Sealed

Crack Sealed

Floor Drain

Piping

Other

—— Interior Walls

—— Main Walls

Floor Type

Area Not Investigated

Carpet

Concrete

Croncrete with Liner

Tile

Other (add notes)

Buildings

DEVELOPING RECREATIONAL MOBILE CROWD-SOURCING APPLICATIONS FOR NEW YORK

Stone Environmental, Montpelier, Vermont, USA
By Katie Budreski, Roger Branon-Rodriguez, Alan Hammersmith, Jeff Herter, and Alex Kuttesch

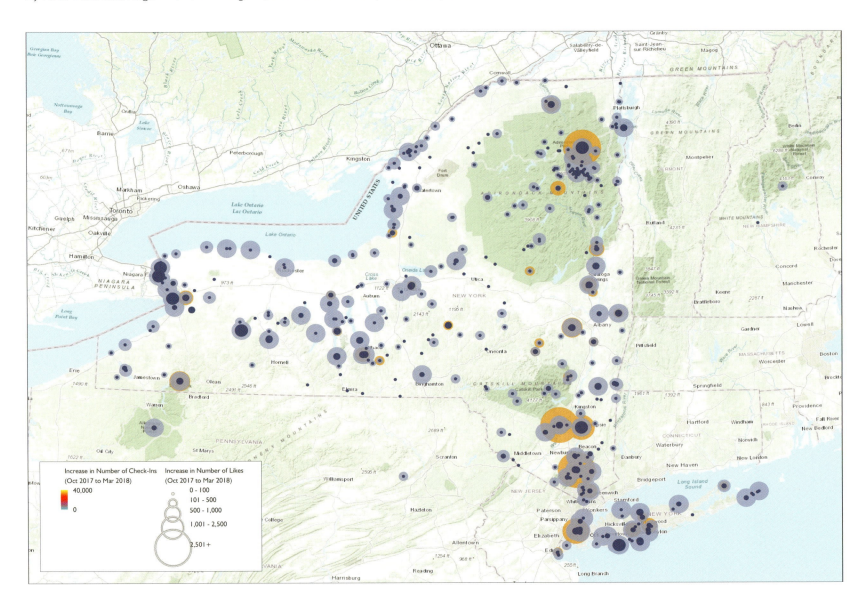

The New York Department of State (DOS) Office of Planning and Development (OPD), in partnership with Stone Environmental, has developed two applications for gathering recreational use and conditions data from recreators in the state of New York. These applications provide OPD with historically difficult-to-capture recreational use data that can be incorporated into offshore, coastal, and inland decision-making and planning efforts.

CONTACT
Katie Budreski
kbudreski@stone-env.com

SOFTWARE
ArcGIS Pro, Adobe Illustrator

DATA SOURCES
Stone, Esri, New York DOS

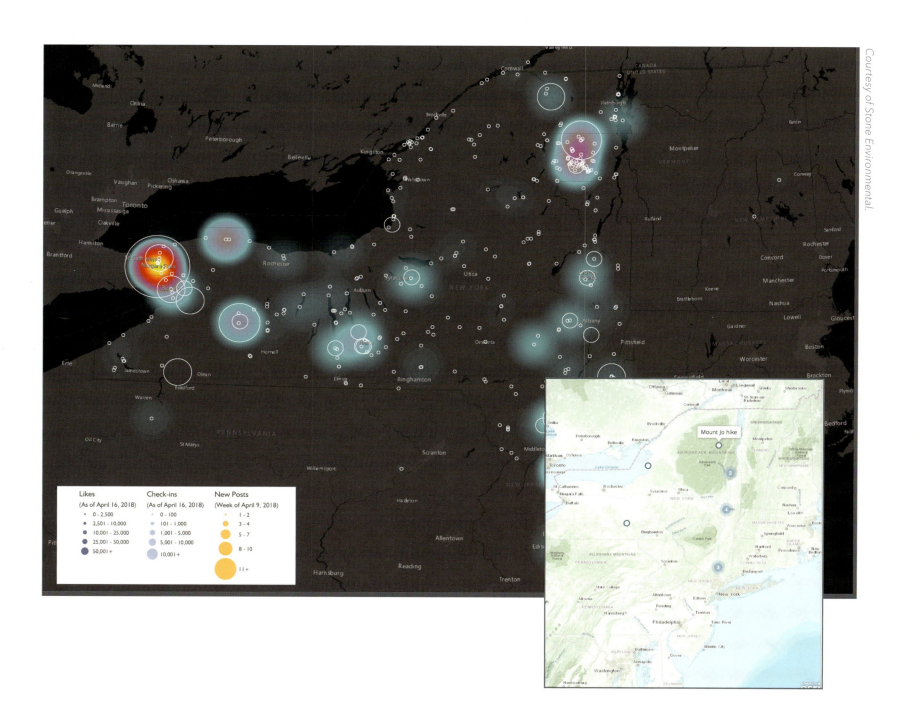

Likes
(As of April 16, 2018)

- 0 - 2,500
- 2,501 - 10,000
- 10,001 - 25,000
- 25,001 - 50,000
- 50,001 +

Check-ins
(As of April 16, 2018)

- 0 - 100
- 101 - 1,000
- 1,001 - 5,000
- 5,001 - 10,000
- 10,001 +

New Posts
(Week of April 9, 2018)

- 1 - 2
- 3 - 4
- 5 - 7
- 8 - 10
- 11+

Mount Jo hike

THE GEODESIGNER'S TOOLBOX

Houseal Lavigne Associates, LLC
Chicago, Illinois, USA
By Devin Lavigne, Nik Davis, and Trisha Stevens

As technology continues to improve and become more accessible, so too must planners diversify the tools they use to develop, conceptualize, and present planning documents. This geodesign toolbox reviews the various programs, applications, software, and even hardware used by Houseal Lavigne Associates. Together, these various resources help to create plans and graphics that are easily legible, appealing, and able to visually communicate important information.

While the traditional "toolbox" features individual tools that are pulled out, used, and returned to their place, the Geodesigner's Toolbox is presented as a fluid wheel. As planning concepts are formed, they travel along the wheel, moving through individual programs and applications in their development. As a result, the tools all work as a single part of the continuous whole by which information is shared, analyzed, conceptualized, and ultimately, visualized.

At its foundation, the Geodesign's Toolbox aims to do two things: inform planning with the data, analysis, public engagement, and research necessary to create viable and exciting plans; and utilize a graphic approach to planning that allows those plans to tell compelling stories, communicate complex ideas, and engage with those for whom planning will mean real world change.

CONTACT

Devin Lavigne
dlavigne@hlplanning.com

SOFTWARE

ArcGIS Pro, ArcGIS Desktop, Business Analyst, CityEngine, GeoPlanner

DATA SOURCES

DuPage County & Village of Downers Grove, Illinois; City of Oshkosh, Wisconsin; City of Bentonville, Arkansas; City of St. Cloud, Minnesota; City of Aurora, Colorado; City of Flint, Michigan; City of Chicago, Illinois; City of Eden Prairie, Minnesota; Esri; Village of Maywood & Cook County, Illinois; ArcGIS Online

Courtesy of Houseal Lavigne Associates, LLC.

IMAGINE FLINT: GEODESIGN IN PRACTICE

Houseal Lavigne Associates, LLC
Chicago, Illinois, USA
By Devin Lavigne, Nik Davis, Brandon Nolin, and John Houseal

Adopted in 2013, the Imagine Flint Master Plan is a prime example of geodesign in practice, emphasizing the importance of geospatial analysis and visually compelling maps and graphics. The incorporation of spatial data throughout the plan helped to not only identify but also clearly communicate the prominent issues Flint faces and the best solutions to address them.

At the heart of the plan is Imagine Flint's placed-based land use map, which shifts focus away from "specific uses" and instead emphasizes the overall character and purpose of different districts within Flint. Each of the plan's maps were created with ArcMap and Adobe Illustrator, and 3D illustrations and CityEngine helped create detailed visualizations of each place type. The plan establishes a vision for each of Flint's twelve distinct areas, supported by considerations for land use, built form, accessibility, and other key factors.

CONTACT
Devin Lavigne
dlavigne@hlplanning.com

SOFTWARE
ArcGIS Desktop, ArcGIS® Business Analyst™, Adobe Illustrator, Esri CityEngine

DATA SOURCES
Houseal Lavigne Associates; Genesee County, Michigan, City of Flint

Courtesy of Houseal Lavigne Associates, LLC.

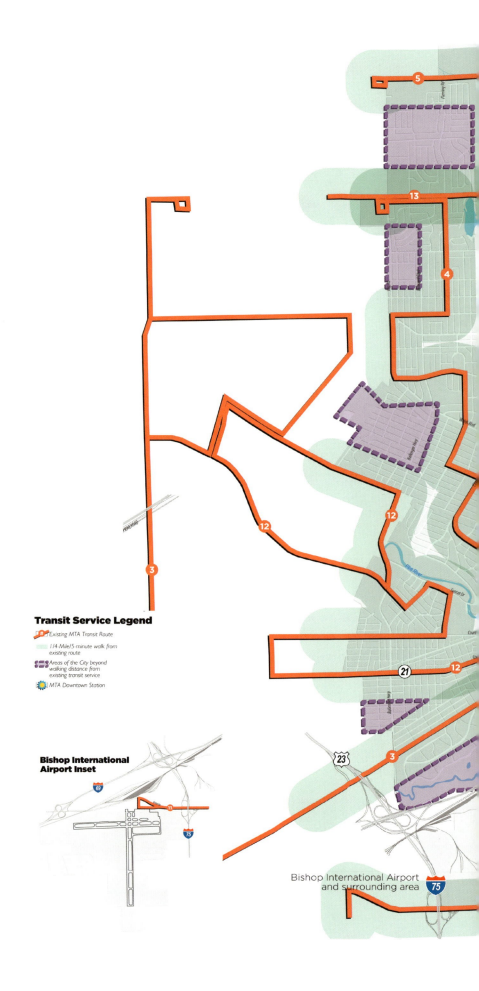

Transit Service Legend

Existing MTA Transit Route

1/4-Mile/5-minute walk from existing route

Areas of the City beyond walking distance from existing transit service

MTA Downtown Station

Bishop International Airport Inset

Bishop International Airport and surrounding area

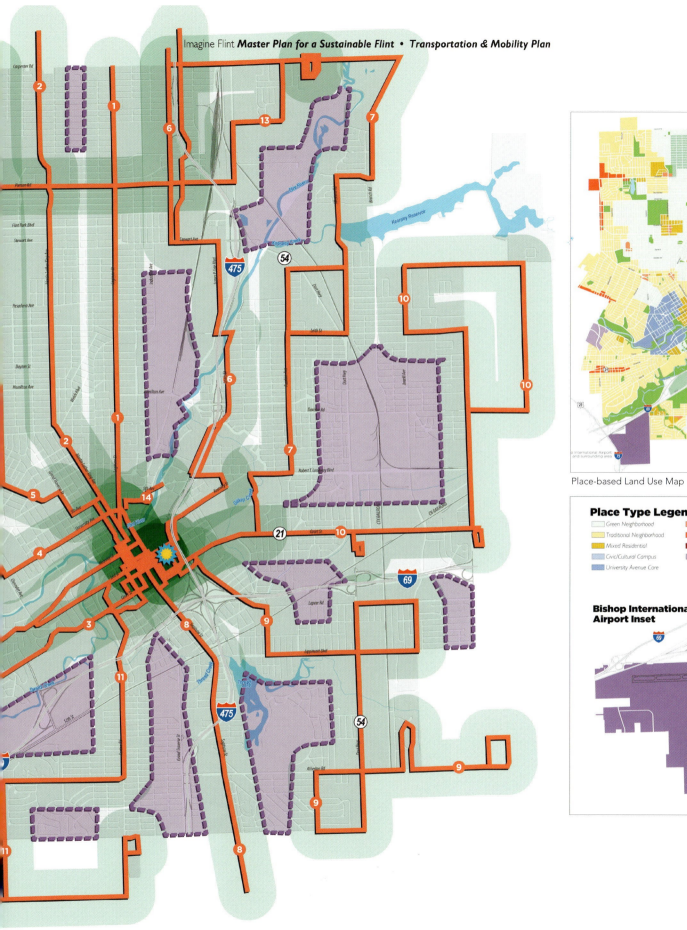

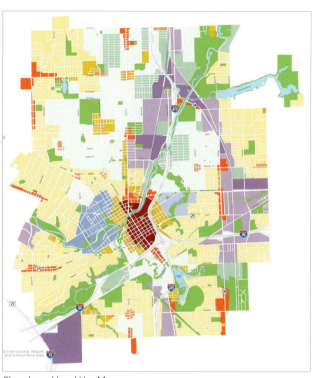

Place-based Land Use Map

Place Type Legend

Green Neighborhood	Neighborhood Center	Production Center
Traditional Neighborhood	City Corridor	Green Innovation
Mixed Residential	Downtown District	Community Open Space & Recreation
Civic/Cultural Campus	Commerce & Employment Center	
University Avenue Core		

Bishop International Airport Inset

REIMAGINING A CITY WITH GIS

Houseal Lavigne Associates, LLC
Chicago, Illinois, USA
By Devin Lavigne

The city of Battle Creek is a regional center within western Michigan, home to global employers, historic neighborhoods, and cultural amenities which make it an inviting community to call home. Recently, the city adopted a new master plan to serve as the city's official guide for land use and development over the next ten to twenty years. It acts as Battle Creek's "playbook," detailing a long-term vision and policy agenda for important issues like land use, housing, parks, infrastructure, transportation, and more. Ultimately, the plan answers the question, what should Battle Creek look like in ten to twenty years, and how do we get there?

Building on the recommendations from the master plan, CityEngine was used to provide a visual understanding of how the city might develop and redevelop as a result of the new plan's recommendations. This includes land use and activity, urban form, and connectivity for ten distinct land use place types, guided by the redevelopment of parcels within the city.

CONTACT
Devin Lavigne
dlavigne@hlplanning.com

SOFTWARE
ArcGIS Pro, Esri CityEngine

DATA SOURCES
Houseal Lavigne Associates; Calhoun County, Michigan, City of Battle Creek

Courtesy of Houseal Lavigne Associates, LLC.

Street Typologies

4-Lane Highway
Center Median
No Sidewalks

5-Lane Arterial
Center Turn Lane
No Sidewalks

4-Lane Arterial
Turn Lane at Intersections
Parkway & Sidewalk

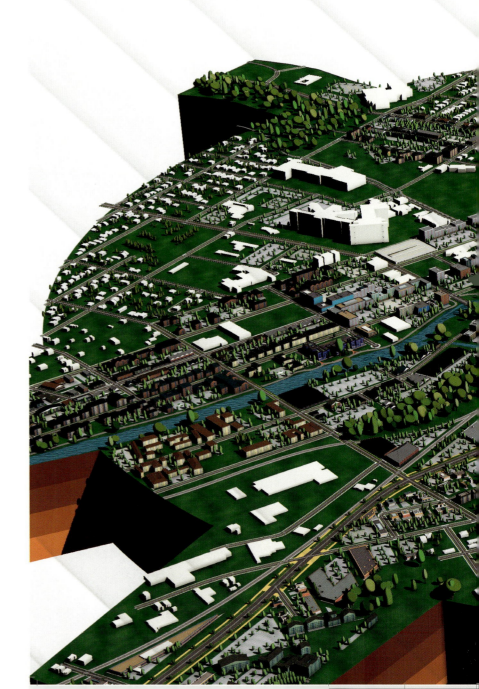

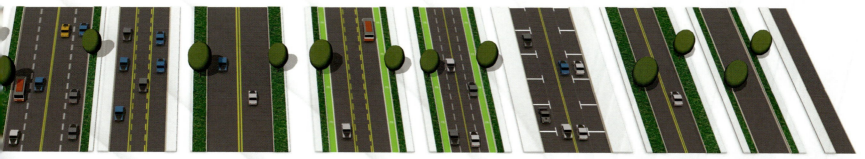

4-Lane Arterial
No Turn Lanes
Parkway & Sidewalk

3-Lane Collector
Center Turn Lane
Wide Sidewalks

2-Lane Collector
Wide Parkway
Narrow Sidewalks

3-Lane Collector
Center Turn Lane
Bike Lanes

4-Lane Collector
Turn Lane at Intersections
Bike Lanes

2-Lane Collector
On-Street Parking
Wide Sidewalks

2-Lane Minor Collector
Wide Parkway
Sidewalks

2-Lane Local
Parkway
Sidewalks

On Ramp

APPLYING SHADOW ANALYSIS THROUGH ESRI CITYENGINE TO PARK LOCATIONS IN DUBAI, UAE

United Arab Emirates University
Al Ain City, UAEU
By Dr. Khaula Alkaabi, Nasir Al Hassan, and Ali Hussien

South Ridge is a cluster of six residential towers located in downtown Dubai. It has six parks beside the towers. GIS techniques (Esri CityEngine) were applied to explore the effects of building shadow over the parks during different periods of time. Such techniques are useful for smart planning and decision making.

A model was developed through 3D GIS techniques to help landscape planners decide where to plant certain plant species to ensure enough sunlight exposure, or where to place park benches so they get enough shade in summer. This effort also supports Dubai Masterplan 2020 to promote sustainable planning, which largely relies on effective and smart urban use arrangements including building, parks, schools, and airports.

CONTACT

Khaula Alkaabi
khaula.alkaabi@uaeu.ac.ae

SOFTWARE

ArcGIS Desktop

Courtesy of United Arab Emirates University.

1. Create new models
2. The sun points locations at different times

10 a.m.

11 a.m.

12 p.m.

1 p.m.

3. The shadow volumes for 10 a.m., 11 a.m., 12 p.m., and 1 p.m.

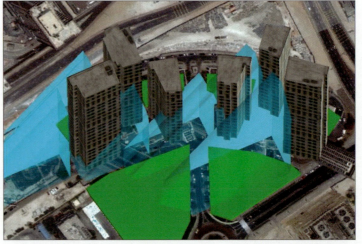

4. Areas where the parks are covered by shadows at 10 a.m.

Shadow Map for Park 1

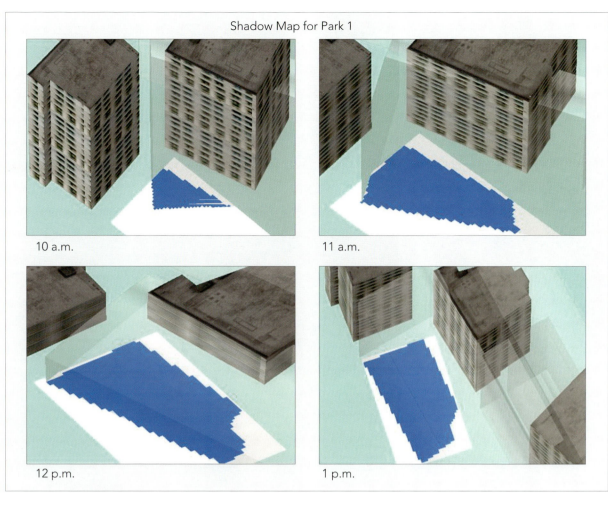

10 a.m.

11 a.m.

12 p.m.

1 p.m.

Shadow Map for Park 2

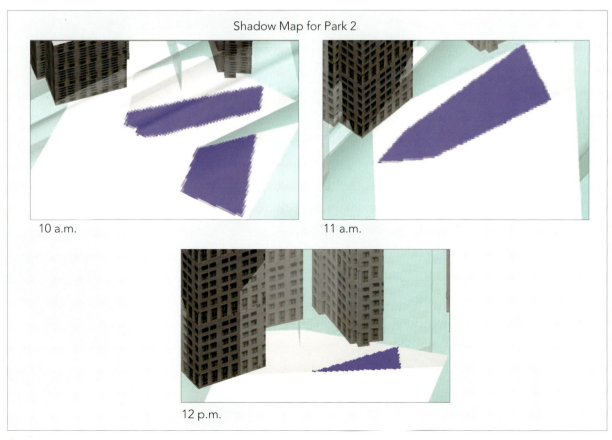

10 a.m.

11 a.m.

12 p.m.

3D ZONING APP

Esri Indonesia
Jakarta Selatan, Indonesia
By Lusiana Suwandi

This app is a prototype created for Dinas Cipta Karya, Tata Ruang Dan Pertanahan DKI Jakarta (Cipta Karya Office, Spatial Planning and Land Agency of DKI Jakarta) to show the future building mass by zoning. They can see which building violates the permit (e.g., exceeds max height) and the amount of allocation for each zoning area.

CONTACT
Lusiana Suwandi
lsuwandi@esriindonesia.co.id

SOFTWARE
ArcGIS Pro, Esri CityEngine

DATA SOURCES
Dinas Cipta Karya, Tata Ruang Dan Pertanahan DKI Jakarta

Courtesy of Esri, HERE, Garmin, FAO, NOAA, USGS, © OpenStreetMap contributors, and the GIS User Community.

SAN DIEGO REGIONAL BIKE MAP

San Diego Association of Governments
San Diego, California, USA
By Rebecca Grover

The San Diego Regional Bike Map is a free resource provided by SANDAG (the San Diego Association of Governments) to help educate riders throughout the region and encourage use of bike facilities. This map displays existing (as of April 2018) bikeways by facility type classification, in addition to identifying steep routes, locations for secure bike parking, and major rail transit. Collaboration between SANDAG and local jurisdictions throughout the county was necessary to identify existing bike facilities for the 2018 map, which previously had not been updated since 2015.

Another key component of this map is the supplemental riding tips and safety information for bicyclists to ensure a comfortable and safe journey for both riders and the vehicles around them.

CONTACT

Rebecca Grover
Rebecca.Grover@sandag.org

SOFTWARE

ArcGISDesktop 10.3 and 10.5, Adobe Illustrator

DATA SOURCES

SANDAG, San Diego Geographic Information Source, Transportation Demand Management iCommute, local jurisdictions of San Diego County

The author would like to acknowledge SANDAG's TDM iCommute, Creative Services, and Active Transportation Planning departments for their role in the creation of this map.

Bikeways in the San Diego Region

multi-use path
A completely separate path for shared use by bike riders, pedestrians, and other non-motorized users with minimal vehicle crossings. Some paths may have restricted access or speed limits.

bike lane
A striped lane for one-way bike travel on a street or highway.

bike route
A shared right-of-way designated by signs only, where bike riders share the roadway with motor vehicles. Also includes streets with "sharrows" or shared lane markings.

freeway shoulder bike access
Some freeway shoulders are open to bike riders. Use of freeway shoulders by inexperienced bike riders is not recommended. Obey all regulatory signs and exit the freeway when required.

other suggested routes
These suggested routes provide additional connections and are not official bikeways. Bike riders should use caution in choosing routes appropriate for their skills and equipment.

≫ ≫ ≫ steep routes
These are bikeways with steep sections that may be difficult for some bike riders. Arrows point uphill.

🔒 Secure bike parking, located at a transit station or Park & Ride lot

┼┼┼┼┼┼┼ Trolley, COASTER, and SPRINTER lines

⚙ San Diego Velodrome

For an interactive version of this map, including routes coming soon, visit 511sd.com/bikemap.

Use the ferry to cross San Diego Bay. Riding bicycles is not allowed on the Coronado Bridge.

SANTEE

EL CAJON

LA MESA

LEMON GROVE

SAN DIEGO

NATIONAL CITY

SDSU

Lake Murray

Sweetwater Reservoir

San Diego County

UNITED STATES

MEXICO

141

NATIONAL RIVERS AND TRAILS OF CALIFORNIA

USDA Forest Service
Vallejo, California, USA
By Daniel Spring

In 1968, both The National Trails System Act and The Wild and Scenic Rivers Act were signed into law by President Johnson. This map celebrates the fiftieth anniversary of the enactment of these two laws by illustrating the multitude of trails and rivers that have since been designated in California for preservation and protection under the two acts.

The National Trails Act and its amendments established three types of national trails that could be designated either by Congress (national scenic and national historic trails) or by the secretaries of the Interior and of Agriculture (national recreation trails). On this map, designated trails are shown as separate line styles based on their designation type, and in the case of historic, each trail is assigned a different color to distinguish them more clearly from each other.

The Wild and Scenic Rivers Act created the National Wild and Scenic Rivers System (NWSRS) and provided processes for either Congress or the Secretary of Interior to designate rivers as part of the NWSRS. For clarity and generalization purposes, only the designated river corridors are shown and labeled on this map, not the segments and classifications.

CONTACT
Daniel Spring
danieljspring@fs.fed.us

SOFTWARE
ArcGIS, Adobe Illustrator, Adobe Photoshop, ArcGIS Desktop 10.1

DATA SOURCES
US Forest Service, Bureau of Land Management, National Park Service, American Trails National Recreation Trails Database

Courtesy of USDA Forest Service.

NATIONAL
Pony Express NRT
Hawley Grade NRT
FOREST
Carson Emigrant NRT

RIVER (LOWER)

Columns of the Giants NRT
NATIONAL
Pinecrest Lake NRT
FOREST

HUMBOLDT
TOIYABE
NATIONAL FOREST
Toiyabe N.F.

TUOLUMNE
OWENS RIVER HEADWATERS

COTTONWOOD CREEK

MERCED
Lewis Creek NRT
Shadow of the Giants NRT
Black Point NRT
Rancheria Falls NRT
Squaw Leap NRT

Methuselah Trail NRT

Lost Lake NRT
KINGS
Kings River NRT
Zumwalt Meadow NRT
Crystal Cave NRT
Congress NRT
Whitney Portal NRT

Summit Trail NRT
Jackass Creek NRT
Cannell Meadow NRT
KERN

Nadeau NRT

AMARGOSA

PACIFIC CREST

Twenty Mule Team NRT

OLD SPANISH

MOJAVE

DESERT

SISQUOC
PIRU CREEK
Santa Cruz/Aliso NRT
SESPE CREEK
Piedra Blanca NRT
Ventura River Parkway NRT
Backbone NRT

High Desert NRT
North Shore NRT
Camp Creek NRT
Sugarloaf NRT
FULLER MILL CREEK
NORTH FORK SAN JACINTO
PALM CANYON CREEK

Silver Moccasin NRT
Gabrielino NRT
West Fork NRT
The Los Angeles River NRT
Mountains to Sea NRT
Aliso Creek Regional Bikeway, Riding and Hiking NRT

BAUTISTA CREEK
Observatory NRT
Inaja NRT
Noble Canyon NRT

Bayside NRT

JUAN BAUTISTA

MEXICO

N

0   10 20 30 40 50 60 70 80 90 100 Miles

0   50   100 Kilometers

1:2,000,000

143

Designed and Produced by USDA Forest Service Pacific Southwest Region,
Information Management, Vallejo, CA in 2018.
Trails and Wild and Scenic Rivers data sources: USFS, BLM, NPS, and American Trails NRT Database.

The USDA Forest Service is an equal opportunity service provider and employer.

SkiAR–SKI RESORT NAVIGATOR

Smart Media Systems, LLC for ROSENGINEERING, JSC
St. Petersburg, Russia
By Ekaterina Gurina and Alena Vasileva

For the ski seasons for 2017 and 2018, we developed the first version of an app for active leisure and sports, SkiAR. The main goal of the app is to navigate the slopes of ski resorts by giving an opportunity to study ski resort maps in 3D and 2D modes. All resort maps can be download to the device memory and work offline.

We made the preconfigured geodatabase and designed the geoprocessing tools and map templates. All ski tracks, cable cars, and lifts inside resorts are connected to the network dataset, so the app can build routes. SkiAR takes into account the user's skiing level, so the user will receive the most suitable routes.

SkiAR also collects statistics about users' behavior at the resort. The next steps of the app's development are related to the implementation of methods for analyzing this collected data and taking into account user's preferences for more useful notifications about resort objects and events. It will help the resort better understand its visitors and perform more segmented analysis to develop its business.

CONTACT
Ekaterina Gurina
ekaterina.gurina@smartms.org

SOFTWARE
ArcGIS Desktop 10.5, SkiAR, Adobe Photoshop CC

DATA SOURCES
Smart Media Systems, digital elevation model data from US Geological Survey, OpenStreetMap

Courtesy of Smart Media Systems, LLC.

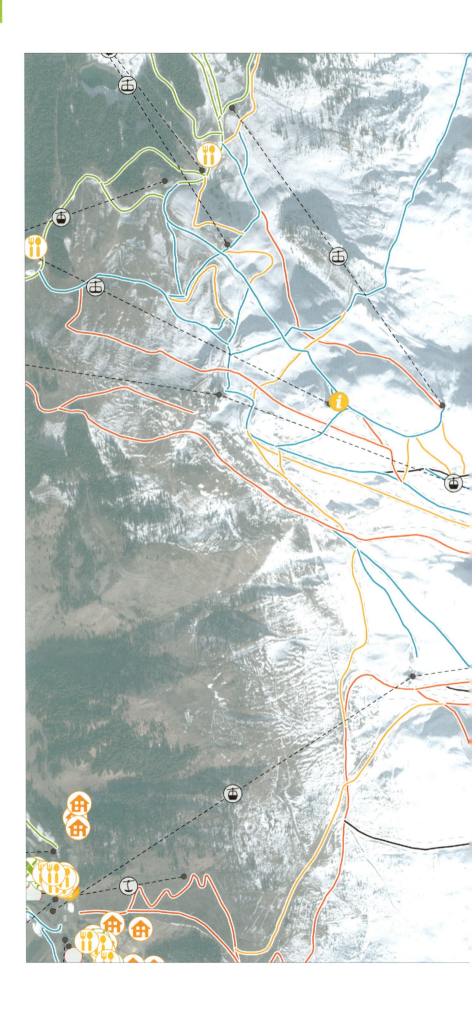

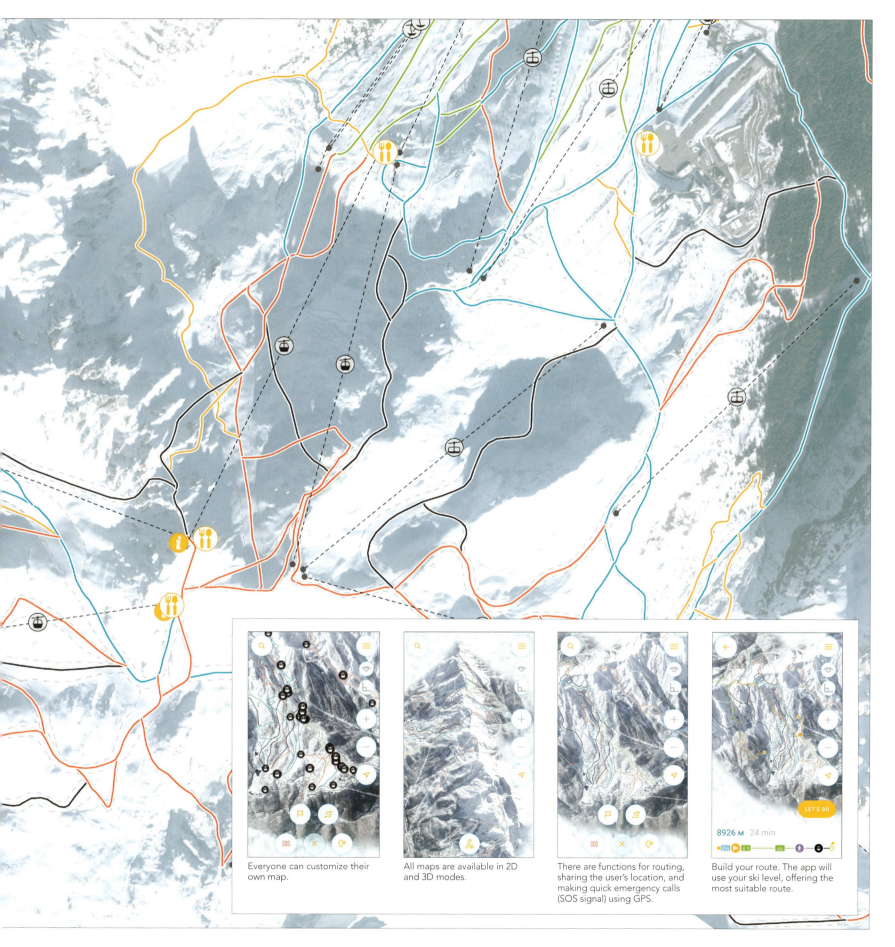

Everyone can customize their own map.

All maps are available in 2D and 3D modes.

There are functions for routing, sharing the user's location, and making quick emergency calls (SOS signal) using GPS.

Build your route. The app will use your ski level, offering the most suitable route.

8926 м 24 min

TOURIST MAP OF NABLUS IN PALESTINE

GSE
Beit Jala, Bethelehem, Palestine
By Michael Younan and Hanna Younan

The Nablus Map is a series of Palestine city maps that Good Shepherd Engineering (GSE) and the Palestine Mapping Center publishes to promote tourism in Palestine. The map is a first-of-its-kind that the Palestine Ministry of Tourism and Antiquities distributes for detailed information of Palestinian cities.

The creation of the map is a team effort by GSE staff, and the information within the map is captured from aerial photography, field work, and descriptive data from the Ministry. The map is distributed by the Ministry free of charge to promote Palestinian heritage and culture, encouraging tourists and visitors to visit the land of Palestine.

CONTACT
Michael Younan
hanna.younan@gse.ps

SOFTWARE
ArcGIS Desktop 10.6

DATA SOURCES
GSE

Courtesy of Good Shepherd Engineering.

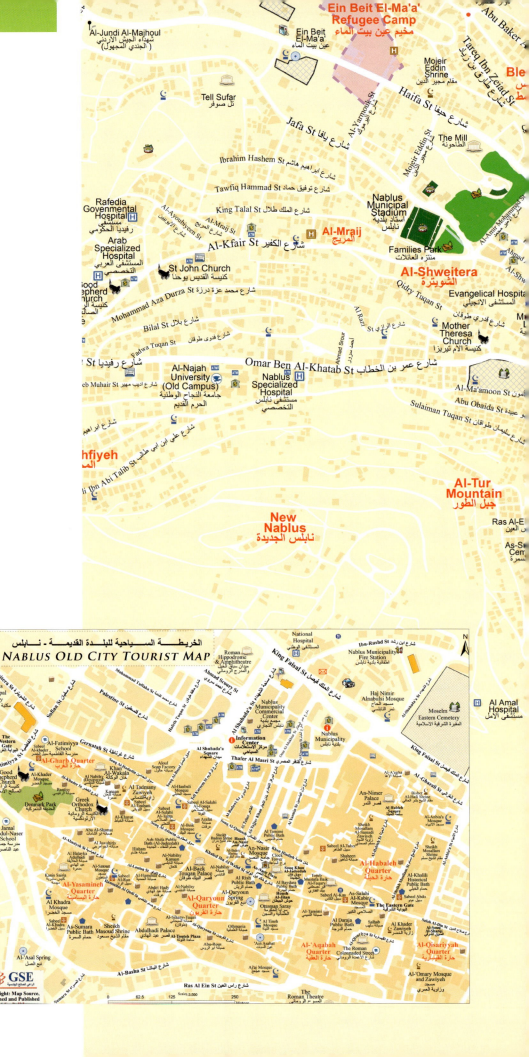

2050 ROAD CONGESTION IN UTAH COUNTY

Mountainland Associaion of Governments

Orem, Utah, USA

By Kory Iman and Tim Hereth

Utah County is the second most populous county in the state of Utah. Over the past few years, Utah County's population growth has exploded and is predicted to continue this trend well into the future. This tremendous growth will have a significant impact on the county's resources now and in the future.

The current transportation infrastructure will also be challenged by this exponential growth. Transportation planners must understand the key issues that will impact travel in the future. This map illustrates locations of future (2050) congestion to assist transportation planners and elected officials in making decisions now that will alleviate these areas of congestion in the future. Sound transportation decisions made today will keep Utah moving well into the future.

CONTACT

Kory Iman

kiman@mountainland.org

SOFTWARE

ArcGIS Desktop 10.5, Adobe Illustrator

DATA SOURCES

Mountainland, Utah County, Utah Automated Geographic Reference Center

Courtesy of Mountainland Association of Governments.

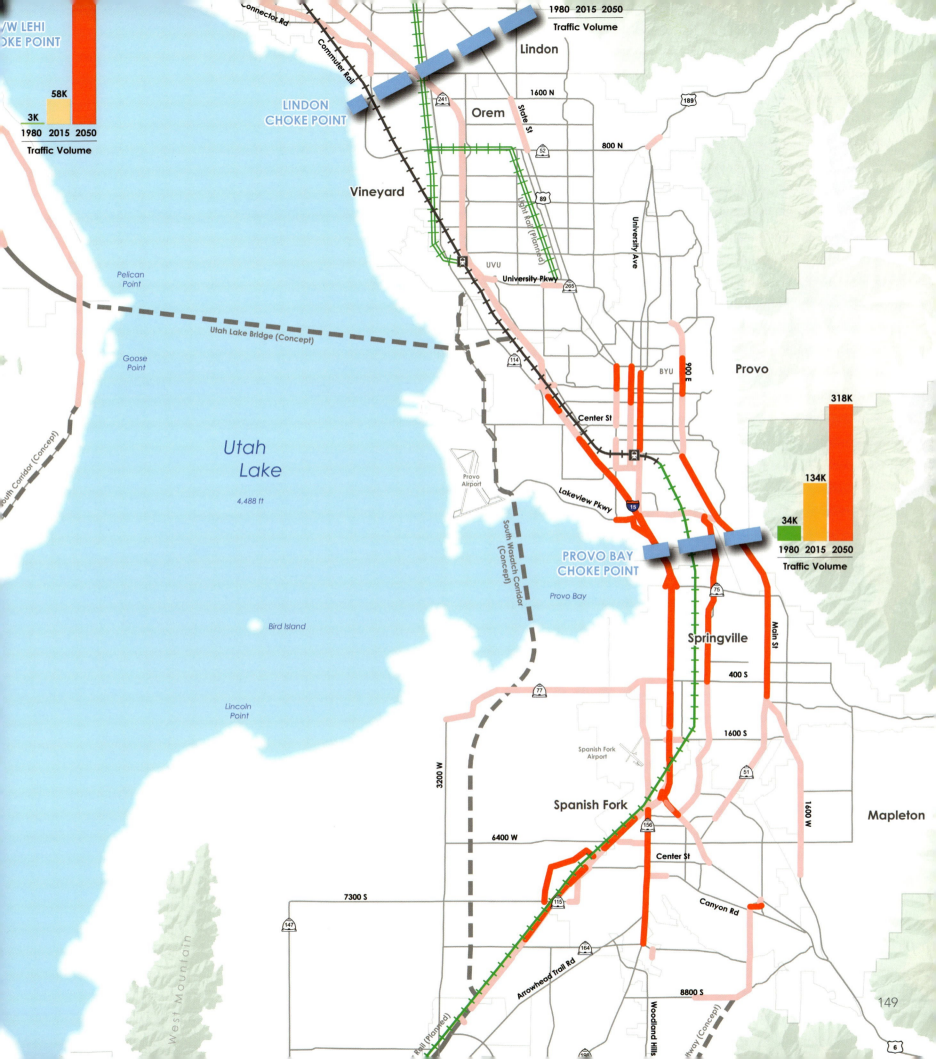

CITY OF LOS ANGELES TRANSIT ORIENTED COMMUNITIES

City of Los Angeles
Los Angeles, California, USA
By Timmy Luong

With the ongoing housing crisis in California affecting people of all ages, a yearn for a call to action was answered within the city of Los Angeles through the passage of Measure JJJ to help provide affordable housing citywide. Pursuant to language within Measure JJJ, Los Angeles Municipal Code (LAMC) 12.22 A.31 was added to create the Transit Oriented Communities (TOCs) Affordable Housing Incentive Program, or TOC Program.

These guidelines will encourage developers to take advantage of a housing density bonus by adding extra affordable housing units within a half-mile radius of a major transit stop. Typically, the argument against increased density is the increased car traffic that accompanies it. However, the TOC program is geared for individuals who apply for affordable housing and will most likely keep using the Metro bus system, thereby minimizing any increase in car traffic.

The following map was created to illustrate the relationship between the Metro bus and rail system along major transit stop corridors. Within the half-mile radius are different tiers of eligibility granted to developers who wish to take advantage of the TOC Program based on the level of service and frequency provided by the Metro system, with Tier 4 being the highest bonus and Tier 1 being the lowest bonus.

CONTACT
Betty Dong
betty.dong@lacity.org

SOFTWARE
ArcGIS Pro 2.1

DATA SOURCES
Southern California Association of Governments, Los Angeles Metro, Metrolink, City of Los Angeles Bureau of Engineering, City of Los Angeles Department of City Planning

Courtesy of Timmy Luong of Los Angeles City Planning.

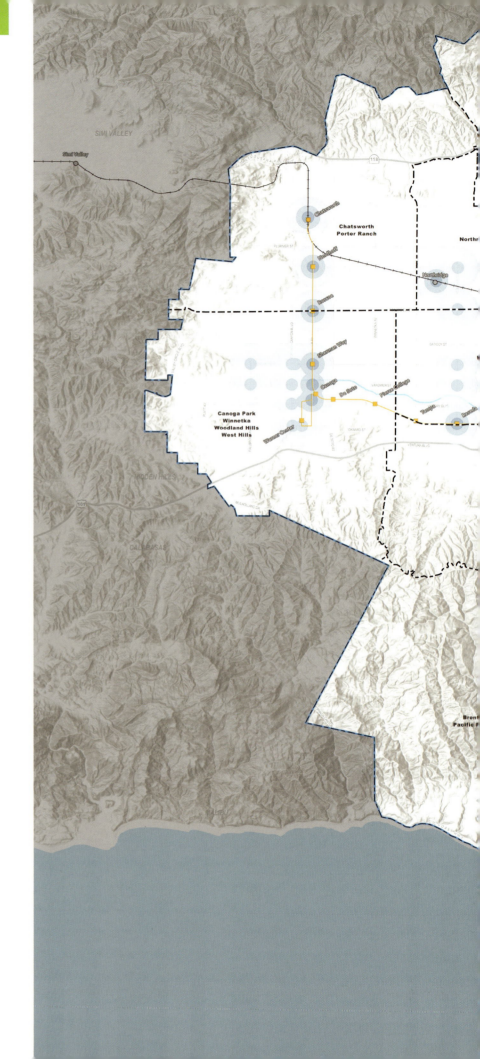

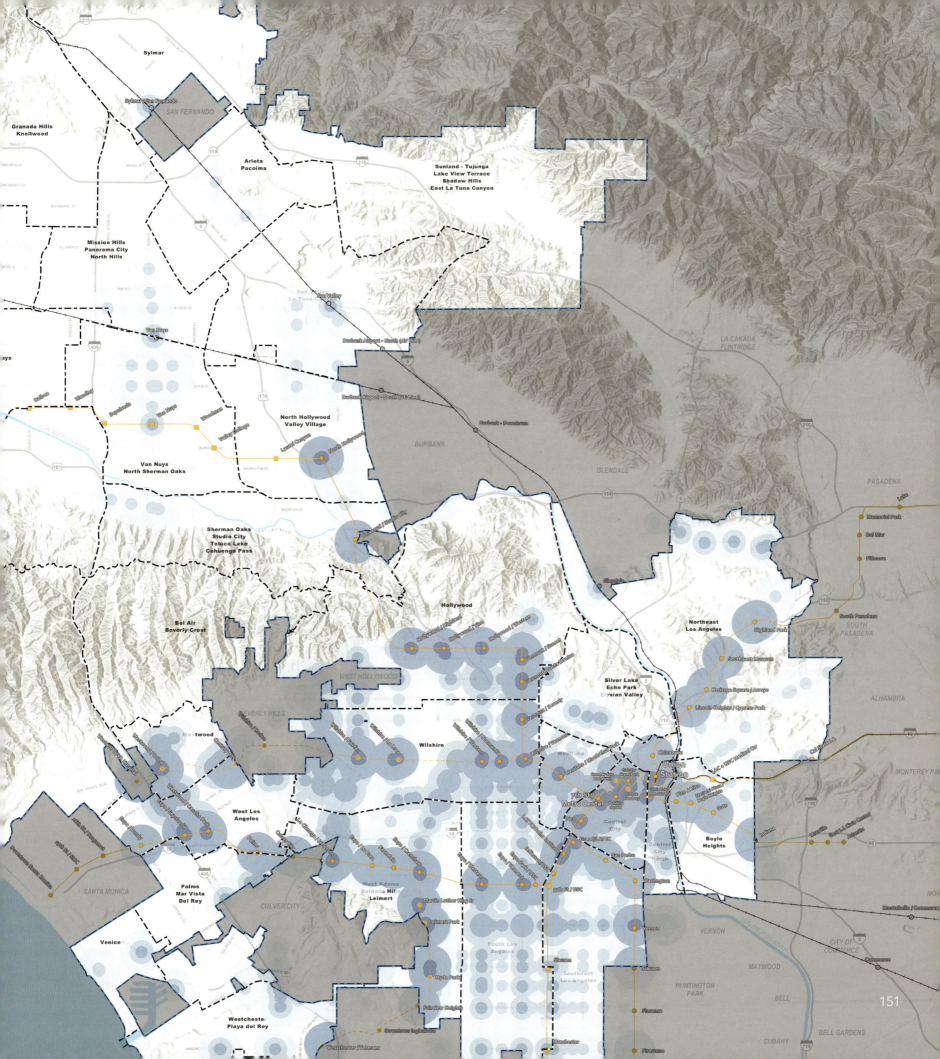

SHARE BICYCLE FOR THE FIRST AND THE LAST MILES ISSUES IN NYC

Tufts University
Beijing, China
By Wencong Xu

We all know that in New York City (NYC), street congestion is one of the biggest problems. Tons of research has been done to improve the ridership of the subway so more room can be made on the streets. One common way to do that is enhancing the accessibility of the subway stations by providing share bicycles. This new kind share bicycle allows people to park them wherever they want because they are not station-based; a user scans a QR code to unlock it.

Our map describes where in NYC we should launch those share bikes so more people would be willing to use them. We believe that by putting them near subway stations and locations in neighborhoods, we can encourage people to travel to the subway stations with share bicycles. We use six factors to evaluate where the suitable locations to launch bikes are, giving a total of 173 places. We believe this map can not only give a hint to the share bicycle companies of where their products might be highly used in NYC but also provide a mode to encourage people to use bicycles and rapid transit.

CONTACT
Wencong Xu
xwc1019@gmail.com

SOFTWARE
ArcGIS Pro 2, ArcGIS Desktop 10.5.1

DATA SOURCES
New York City Open Data, US Census Bureau, New York Citi Bike, Twitter, New York State Open Data

Courtesy of Esri, HERE, DeLorme, MapmyIndia, OpenStreetMap contributors, and the GIS user community, Esri, MapmyIndia.

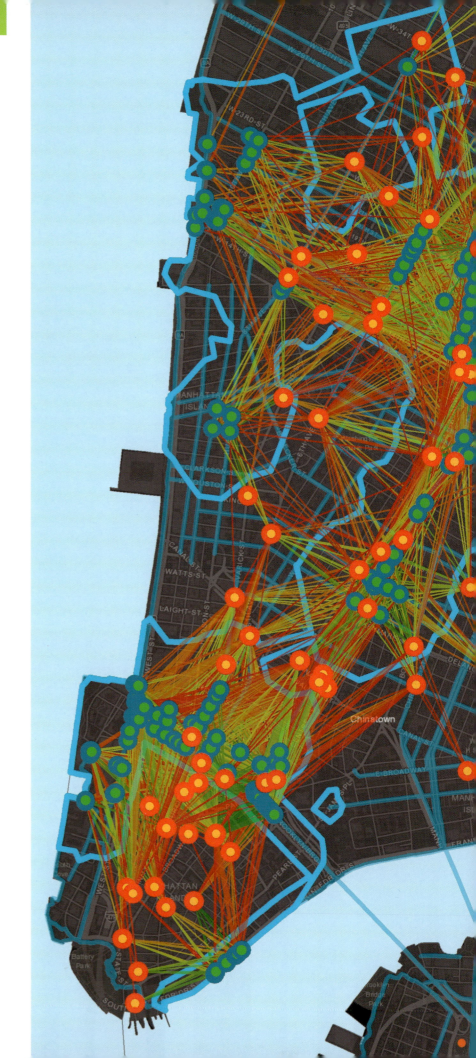

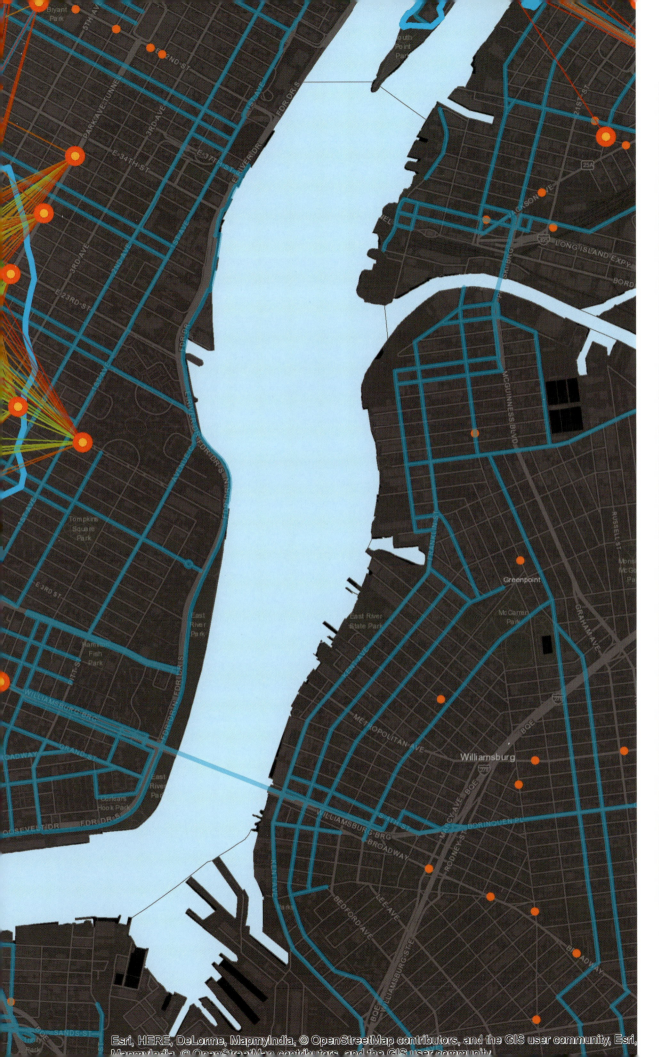

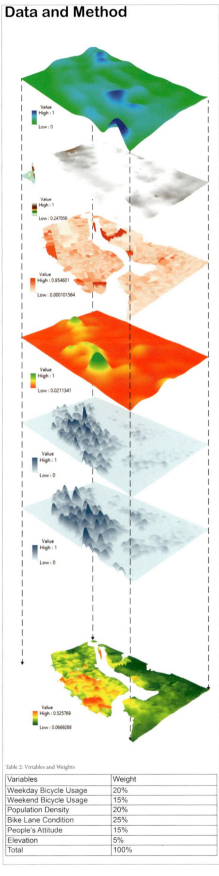

Data and Method

Table 2: Variables and Weights

Variables	Weight
Weekday Bicycle Usage	20%
Weekend Bicycle Usage	15%
Population Density	20%
Bike Lane Condition	25%
People's Attitude	15%
Elevation	5%
Total	100%

icROUTE, TRANSMISSION LINE ROUTING TOOL

HDR Inc
San Diego, California, USA
By Yuying Li and Anders Burvall

The icRoute tool is a web-based application developed to provide a collaborative environment to analyze and help select future transmission line routes. The tool allows users to identify critical impacts, environmental concerns, and other regulatory restrictions for a project and provides different options for preferred routes using a least-cost path algorithm. The project design teams and clients can work collaboratively to evaluate various route options, review the impacts on the fly, and generate quantitative comparison among potential routes.

CONTACT

Anders Burvall
anders.burvall@hdrinc.com

SOFTWARE

ArcGIS Pro, Adobe Illustrator SC3

DATA SOURCES

US Geological Survey, Federal Emergency Management Agency, US Fish and Wildlife Service

Courtesy of HDR.

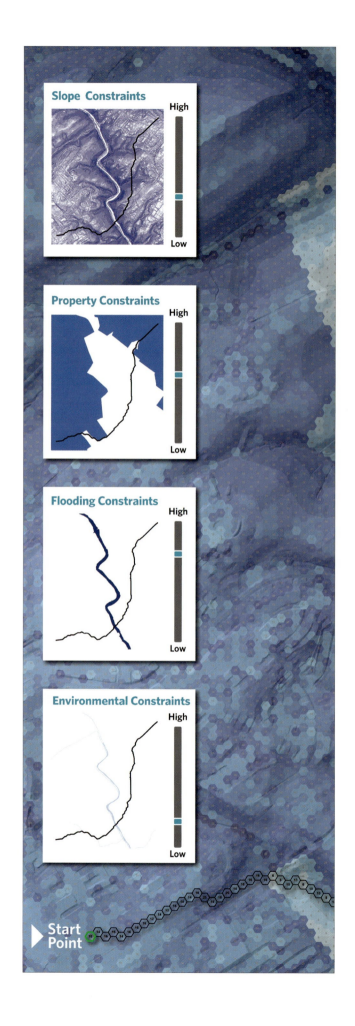

FIGURE DESIGN FOR A REGIONAL LIGHT RAIL FINAL ENVIRONMENTAL IMPACT STATEMENT

HDR Inc
Bellevue, Washington, USA
By Jim Glassley

Public and agency understanding of a complex transit environmental impact statement (EIS) was achieved by creating design and organizational standards early on in the process. With a consistent look and feel to the graphics, readers were able to easily digest the message and information each map was conveying. It also made the project team more efficient by adopting standard data organization structures and mapping templates that enabled consistent and repeatable mapping and analysis.

HDR's GIS program led this effort, and we worked with local municipalities to acquire and organize four jurisdictions' worth of GIS data sets and worked closely with the project design team to analyze nineteen different options of the potential alignment and impacts. We also worked closely with discipline leads to present the results of the multiple alignment options in the Final Environmental Impact Statement. This helped Sound Transit in Washington work through the EIA process more easily and supported better decision-making.

CONTACT
Jim Glassley
jim.glassley@hdrinc.com

SOFTWARE
ArcGIS Desktop 10.4, Adobe Illustrator SC3

DATA SOURCES
King County, Washington; City of Des Moines, Washington; City of Federal Way, Washington; City of Kent, Washington; City of SeaTac, Washington; Washington State Department of Health (2014, 2015, 2016)

Courtesy of HDR, Sound Transit.

Land Use
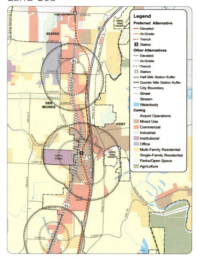

Built Environment

Wellhead Protection
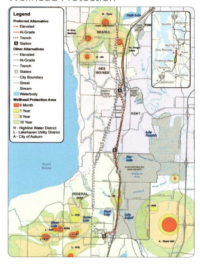

Parks

Right-of-Way
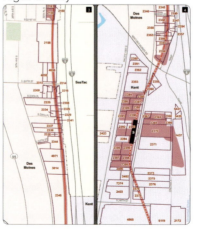

Habitat

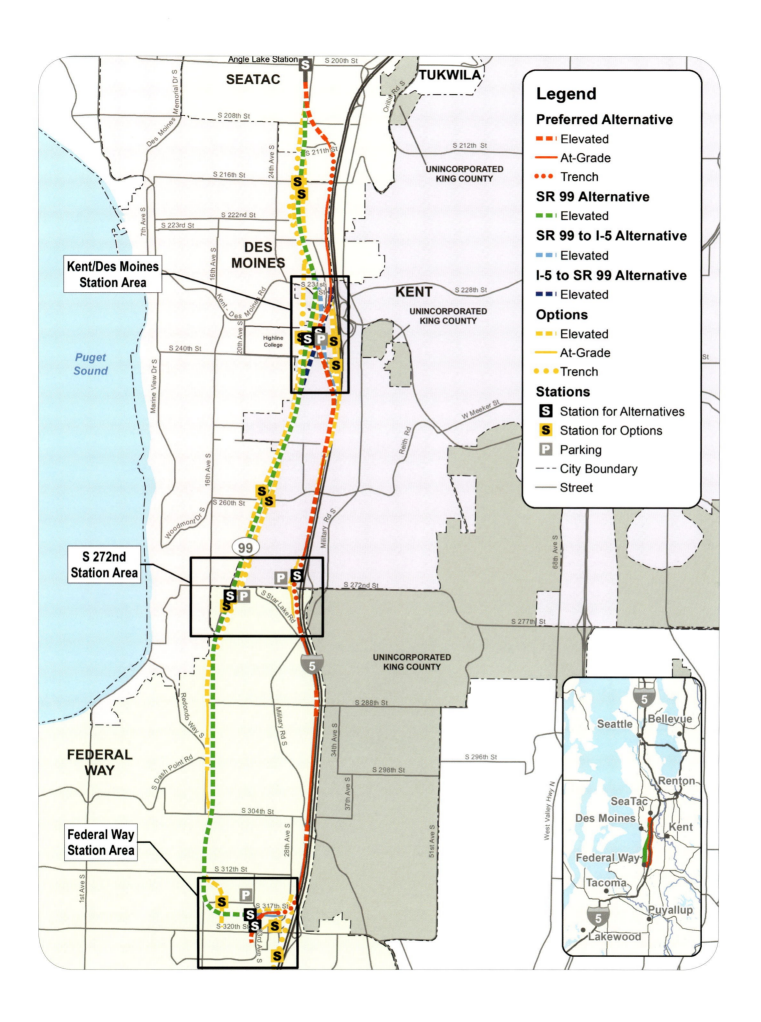

157

OBSTRUCTION IDENTIFICATION SURFACES FOR LAX RUNWAYS

City of LA/LAWA
Los Angeles, California, USA
By Abdel Khineche

The City of Los Angeles GIS Support Services Division at the Los Angeles International Aiport (LAX) has taken up the challenge to create the Airfield Imaginary Surface Areas for LAX using the ArcGIS® for Aviation: Airport extension. They represent the hypothetical surfaces above and around an airfield where there are height restrictions that prohibit obstructions to navigable space and are a requirement of the Federal Aviation Administration.

The models are based on each runway specification and built on criteria specific to the type of airport (civil, military, etc.), as well as the visibility conditions and the type of equipment in place at that airport. The generated surfaces are superimposed over potential ground obstacles that might interfere with flight paths and usually extend far beyond the airport boundaries. For the runways, we used the precision instrument approach, criteria specified by our airfield team at LAX.

CONTACT
Abdel Khineche
akhinech@sbcglobal.net

SOFTWARE
ArcGIS Pro, ArcGIS, Desktop, ArcGIS for Aviation: Airports/Global Mapper

DATA SOURCES
LAX Airport Layout Plan

Copyright of GISSSD/IMTG/LAWA.

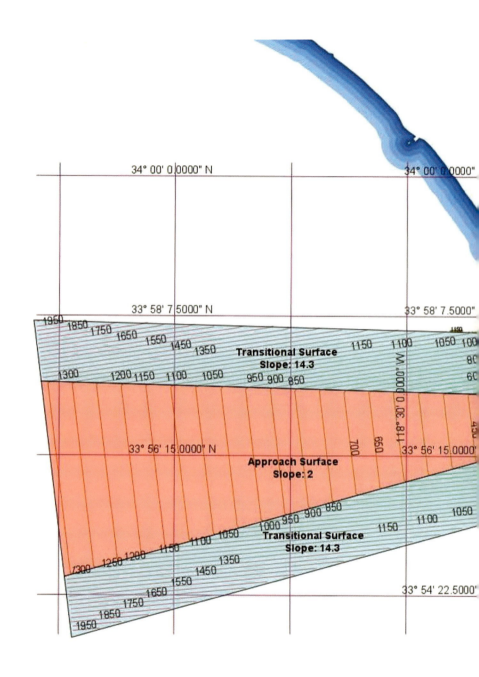

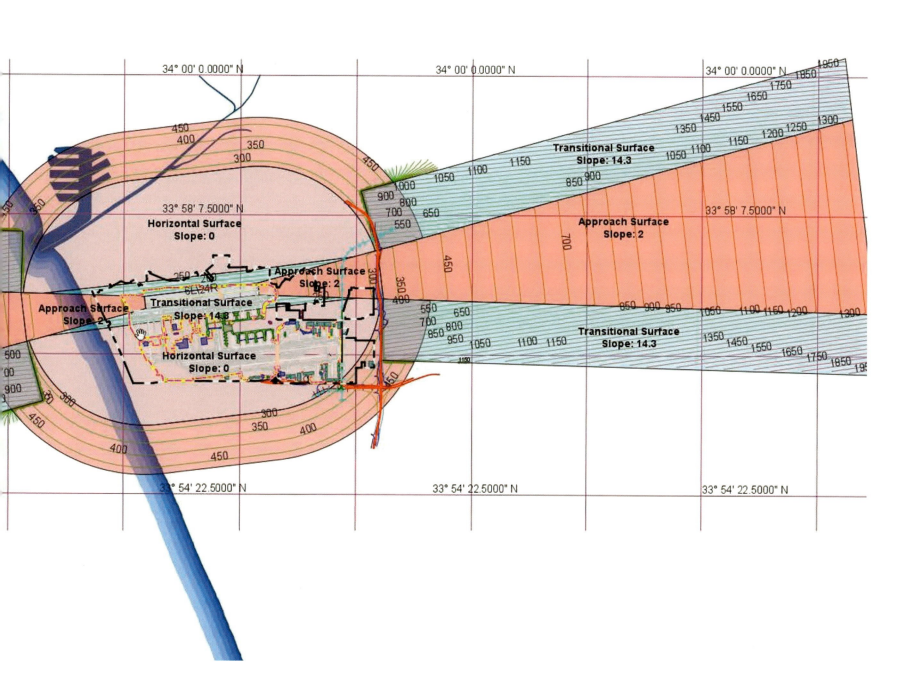

GREATER LONDON IN 3D IN A WEB APPLICATION SHOWING LONDON UNDERGROUND STATIONS

Garsdale Design Limited
Sedbergh, Cumbria, United Kingdom
By Garsdale Design Limited

Three-dimensional building visualizations can be a powerful communication and analytical tool, especially for those in the geodesign professions. Garsdale Design created this 3D London web application, or 3D mApp, using CyberCity3D data in combination with publicly available open data (from Ordnance Survey and the Environment Agency). In addition, some custom 3D modeling in SketchUp was used and combined to create a massive 3D model of Greater London hosted on ArcGIS Online. The 3D basemap of Greater London used in the application allows clients to swap out basemap buildings for their proposals quickly. All of this can be viewed using a web browser (desktop or mobile), or in ArcGIS Pro and ArcGIS Earth.

CONTACT
Elliot Hartley
elliot.hartley@garsdaledesign.co.uk

SOFTWARE
ArcGIS Pro, ArcGIS Online, CityEngine

DATA SOURCES
3D building data from CyberCity3D, tube stations from OpenStreetMap

Contains CyberCity3D 3D building data (2018). London Underground network (2018) © OpenStreetMap contributors. Contains OS data © Crown copyright and database right (2018). Contains Open Government LiDAR data © Environment Agency 2016 which contains public sector information licensed under the open government license v3.0.

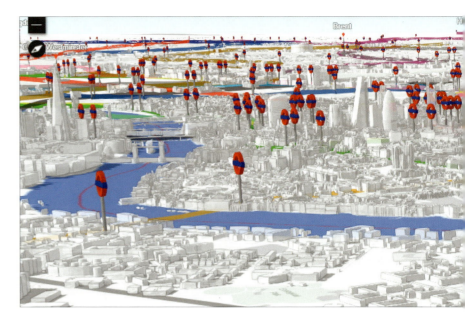

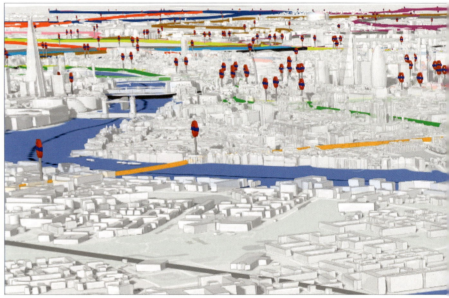

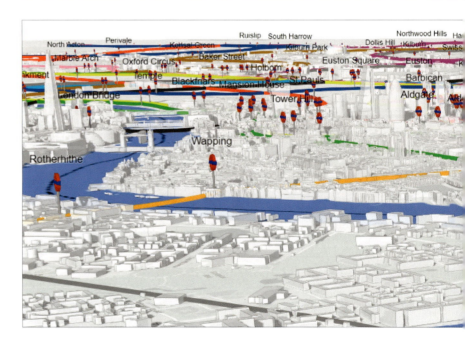

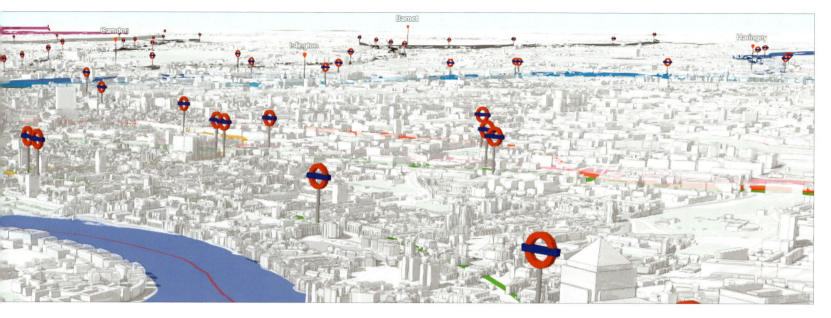

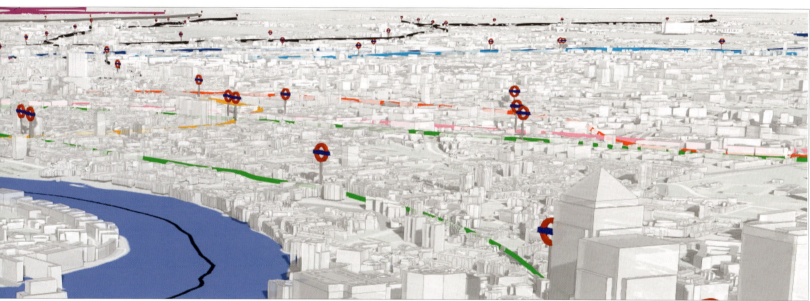

GEO SCHEMATICS–CONDUIT MANAGER FOR EFFECTIVE CAPACITY PLANNING AND LEASING

RMSI North America
New York City, New York, USA
By Amit Rishi

Conduit Manager functionality facilitates access to the precise information on duct alignment and configuration. Coupled with integrated computer-aided design and ArcGIS, the system provides planners accurate information about their formation, position, and vacant duct. Smart tools allow them to perform advance analysis on the collapsed duct details to support informed planning decisions while reducing field visit efforts up to 50 percent. Furthermore, it also helps utility cooperatives/operators to manage accurate billing, increasing revenue generation through leasing out vacant ducts by integrating third-party details into the duct information.

CONTACT
Amit Rishi
amit.rishi@rmsi.com

SOFTWARE
Integrated CAD and Esri GIS systems

DATA SOURCES
RMSI

Courtesy of RMSI North America.

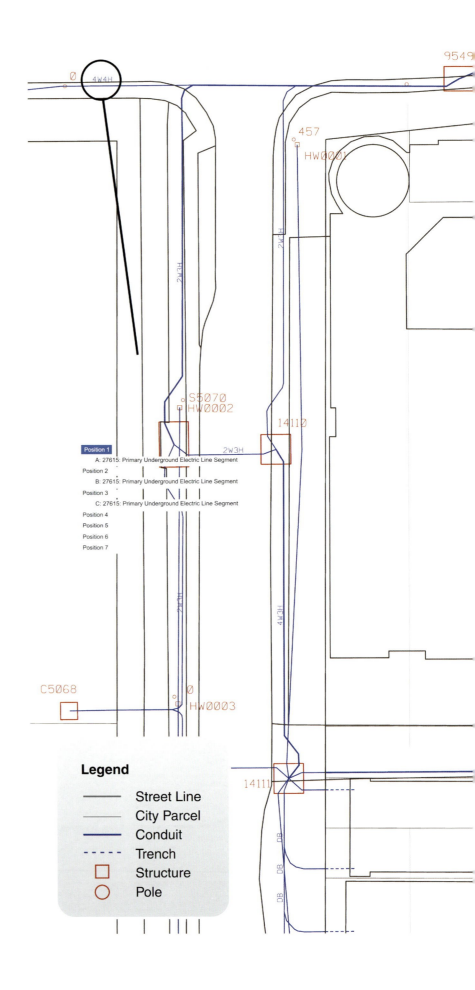

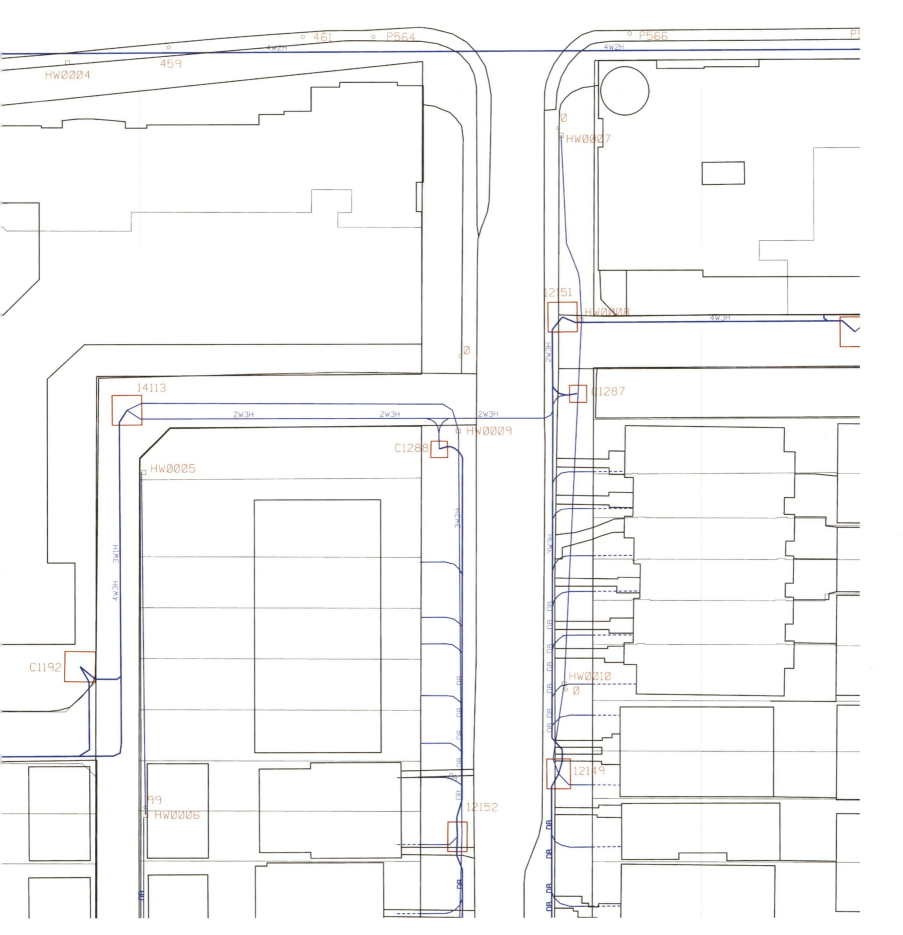

ABU DHABI DEPARTMENT OF ENERGY'S REAL-TIME OPERATIONS DASHBOARDS

Department of Energy
Abu Dhabi, United Arab Emirates
By Khalid Saddique Akbar and Shaima Al Hammadi

The Department of Energy (DOE) in Abu Dhabi, UAE, had some problems to solve—displaying real-time data in a way that provides DOE managers and engineers with insights into their assets and their people on the field. As such, DOE leveraged the latest version of the Operations Dashboard for ArcGIS and integrated it with Supervisory Control and Data Acquisition/Distribution Management System (SCADA/DMS), Oracle Customer Care & Billing System (CC&B), and Collector for ArcGIS to collect and display the location-specific data that form the backbones of these dashboards.

By integrating our ArcGIS/ArcFM GIS platform with the SCADA/DMS and CC&B systems, DOE/Abu Dhabi Water and Electricity Authority is able to visualize, monitor, and prioritize both planned and unplanned incidents and outages in the network along with reporting outage restoration times and the numbers and locations of affected customers while making use of charts, graphs, and queries to get to critical data quickly and effectively.

With real-time load data coming in through OSISoft's PI System, the second dashboard displays asset utilization and how loaded certain substations and transformers in the network are compared to others. A third dashboard to monitor, track, and manage the field crews and their daily progress at the site was developed on top of the Operations Dashboard giving a real-time view of field crew locations, along with the data as they collected it from the field through integration with Collector for ArcGIS.

CONTACT
Khalid Saddique Akbar
ksakbar@adwea.ae

SOFTWARE
Operations Dashboard for ArcGIS Desktop 10.6

DATA SOURCES
SCADA/DMS, Oracle CC&B, OSISoft PI System, Collector for ArcGIS

Courtesy of DOE.

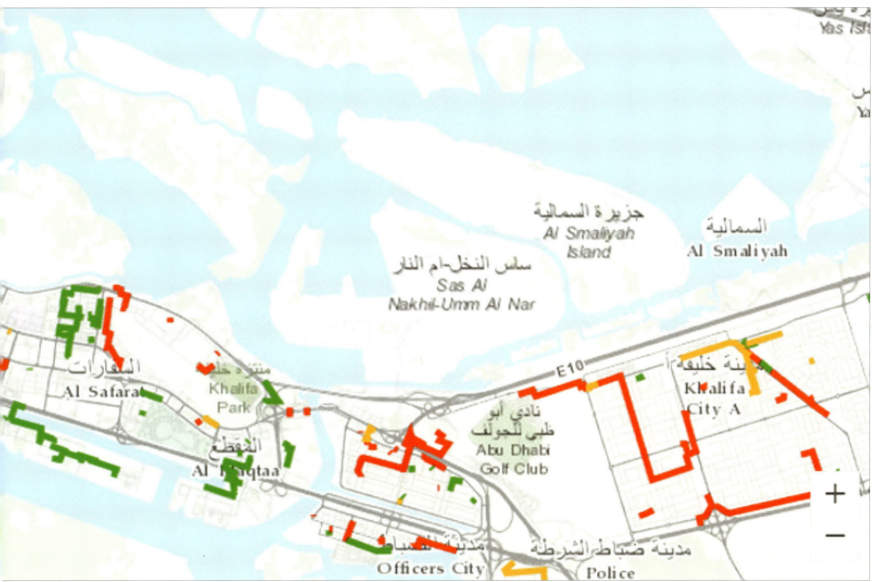

PLUGGING IN BEYOND THE GRID IN TANZANIA: GAUGING SUITABILITY FOR OFF-GRID ELECTRIFICATION

Fletcher School, Tufts University
Boston, Massachusetts, USA
By Matthew Arnold

Nearly one billion people lack consistent electricity access, mostly among rural communities around the world that often remain out of reach of electricity access because of limited grid infrastructure. In Tanzania, the power grid links together urban centers in the country but remains largely undeveloped in rural areas. Given Tanzania's high rate of economic growth and dearth of electricity infrastructure for most of its population, there is heightened potential for off-grid solutions. This opportunity is particularly prominent with micro-hydro, wind, and PV solar technologies. The goal of this project was to determine optimal areas for off-grid electricity potential in Tanzania.

This project involved analyzing four categories of data: hydro potential, wind potential, solar potential, and proximity of population to power substations (the farther away, the less likely a household will be connected to the grid). This data was then aggregated into composite scores to show the combined potential of off-grid technologies and then overlaid with proximity to the existing grid, generating an overall scoring rubric for off-grid suitability across the country.

CONTACT
Matthew Arnold
marnoldw@gmail.com

SOFTWARE
ArcGIS Desktop 10.5.1

DATA SOURCES
Tanzania Electric Supply Company, Tanzania National Bureau of Statistics, World Bank Group, Global Atlas, Esri, DIVA-GIS, US Agency for International Development - Power Africa

Courtesy of the Fletcher School and Tufts University GIS Program.

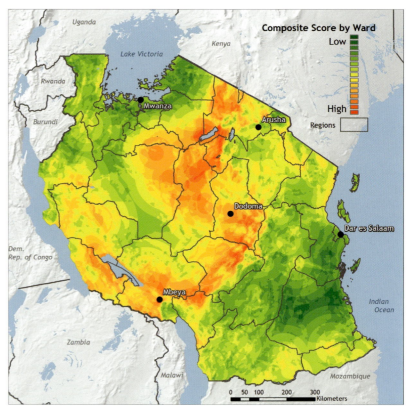

Composite Score of Hydro, Solar, and Wind, by Ward

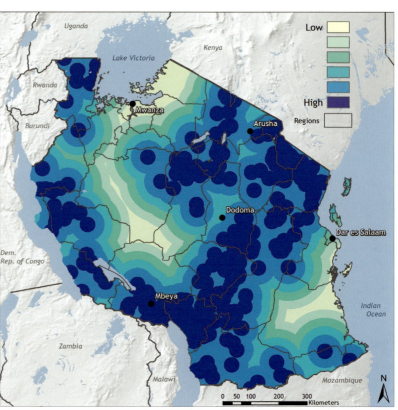

Hydro Potential by Ward

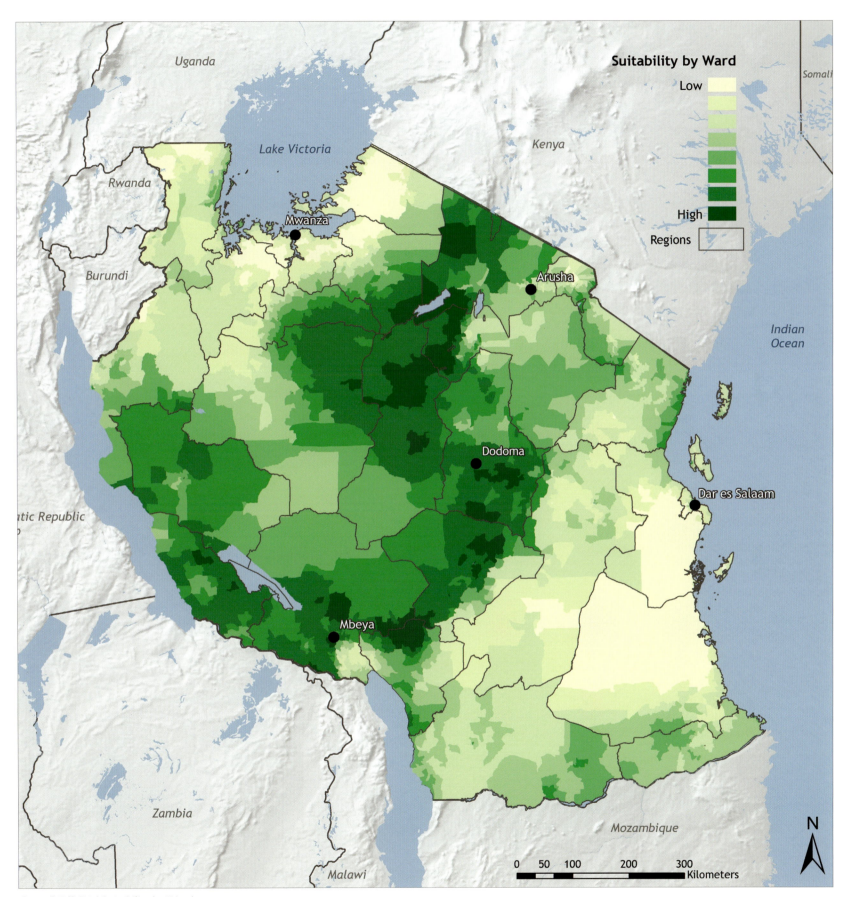

Overall Off-Grid Suitability by Ward

LOWER COLORADO RIVER AUTHORITY ELECTRICAL SUBSTATIONS AT RISK TO WILDLAND FIRE

Austin Community College
Austin, Texas, USA
By Brice Culbertson

The Bastrop County Complex fire was the most destructive wildfire in Texas history. On September 4, 2011, three separate fires which started due to downed power lines merged into one large blaze that burned east of the city of Bastrop. Two people were killed by the fire, which destroyed 1,673 homes and inflicted an estimated $325 million of property damage. The fire underscored the threat wildfires pose to Lower Colorado River Authority infrastructure, especially the numerous substations located in rural and under-developed areas. These maps seek to assess the risk wildfires pose to these substations, which were collected using Esri basemap imagery and heads-up digitizing their locations.

CONTACT

Brice Culbertson
brice.culbertson@g.austincc.edu

SOFTWARE

ArcGIS Desktop 10.5.1

DATA SOURCES

Austin Community College, Texas Natural Resources Information Systems, Capital Area Council of Governments, Texas Wildlife Risk Assessment Portal, Texas A&M Forest Service

Courtesy of Austin Community College.

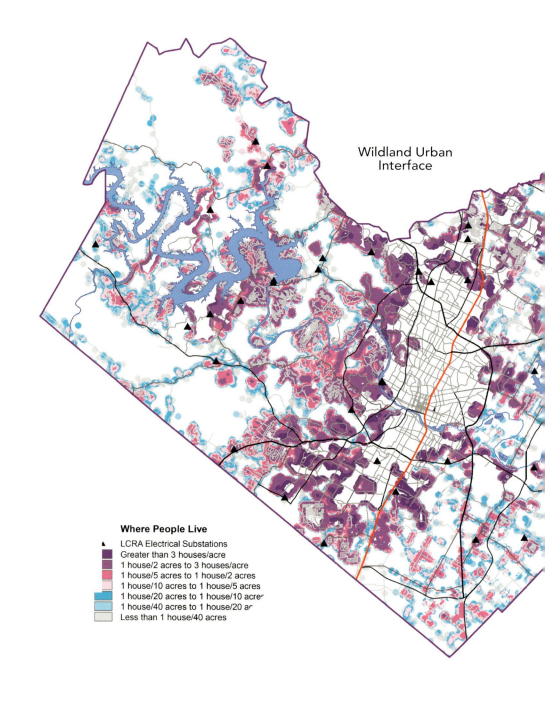

Wildland Urban Interface

Where People Live

▲ LCRA Electrical Substations
 Greater than 3 houses/acre
 1 house/2 acres to 3 houses/acre
 1 house/5 acres to 1 house/2 acres
 1 house/10 acres to 1 house/5 acres
 1 house/20 acres to 1 house/10 acres
 1 house/40 acres to 1 house/20 acres
 Less than 1 house/40 acres

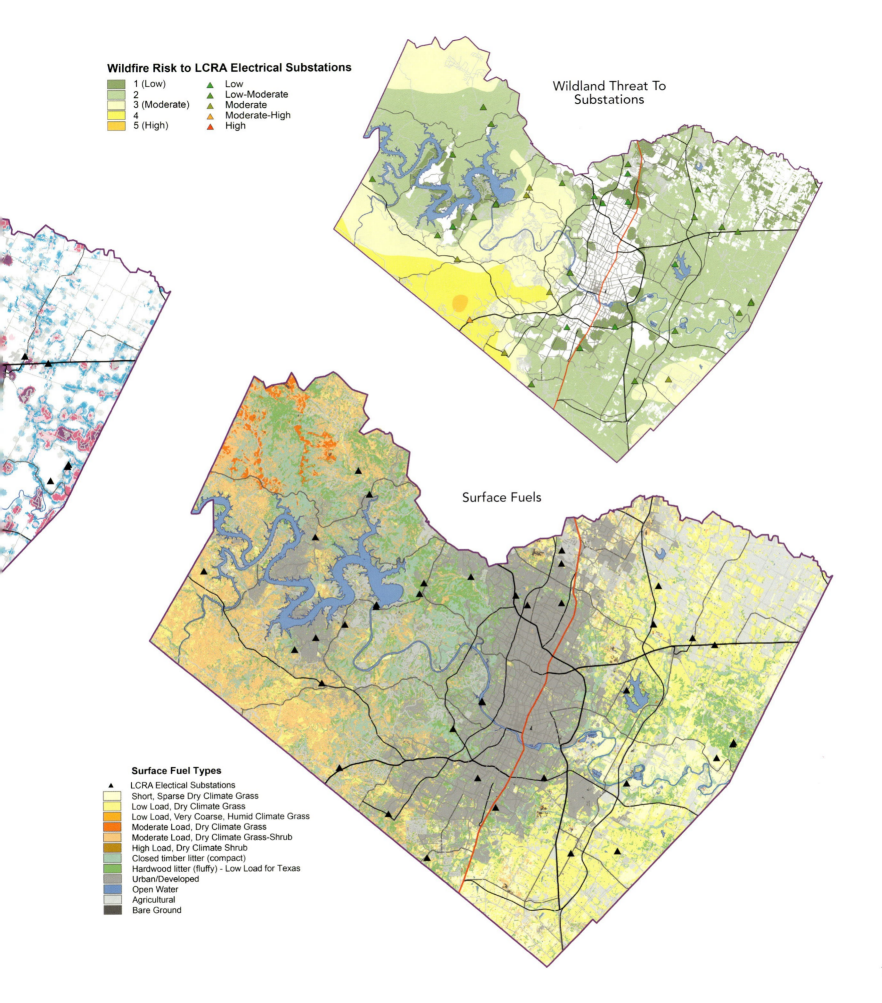

Wildfire Risk to LCRA Electrical Substations

- 1 (Low)
- 2
- 3 (Moderate)
- 4
- 5 (High)

- ▲ Low
- ▲ Low-Moderate
- ▲ Moderate
- ▲ Moderate-High
- ▲ High

Wildland Threat To Substations

Surface Fuels

Surface Fuel Types

- ▲ LCRA Electical Substations
- Short, Sparse Dry Climate Grass
- Low Load, Dry Climate Grass
- Low Load, Very Coarse, Humid Climate Grass
- Moderate Load, Dry Climate Grass
- Moderate Load, Dry Climate Grass-Shrub
- High Load, Dry Climate Shrub
- Closed timber litter (compact)
- Hardwood litter (fluffy) - Low Load for Texas
- Urban/Developed
- Open Water
- Agricultural
- Bare Ground

ELECTRIC NETWORK IN THE CZECH REPUBLIC

HSI, spol. s r.o., Unicorn Group Member
Prague, Czech Republic
By HSI, spol. s r.o., Unicorn Group Member

The mission of CEZ Distribuce, a. s., is to distribute electricity to private individuals and legal entities and to continuously improve quality and reliability of supply to all customers. The company manages distribution grid assets and controls them through a grid control center. The GIS system is one of the critical applications and is used in many processes across the whole company.

The company data is managed in Esri technology and stored in a geodatabase through a web client. The web clients are also based on Esri technology and have several important roles. The most important role is narrow access to geodata of the company for all employees. They can quickly locate the position of the electric network on the map and get other information about network items such as the address or cadastral information.

This map shows different electric networks above topographic maps in a town in Czech Republic. Different colors of the network represent different voltage level.

CONTACT
Ondrej Žák
ondrej.zak@unicorn.com

SOFTWARE
ArcGIS Desktop 10.2.1, ArcGIS Server 10.3

DATA SOURCES
ČEZ Distribuce, a. s.

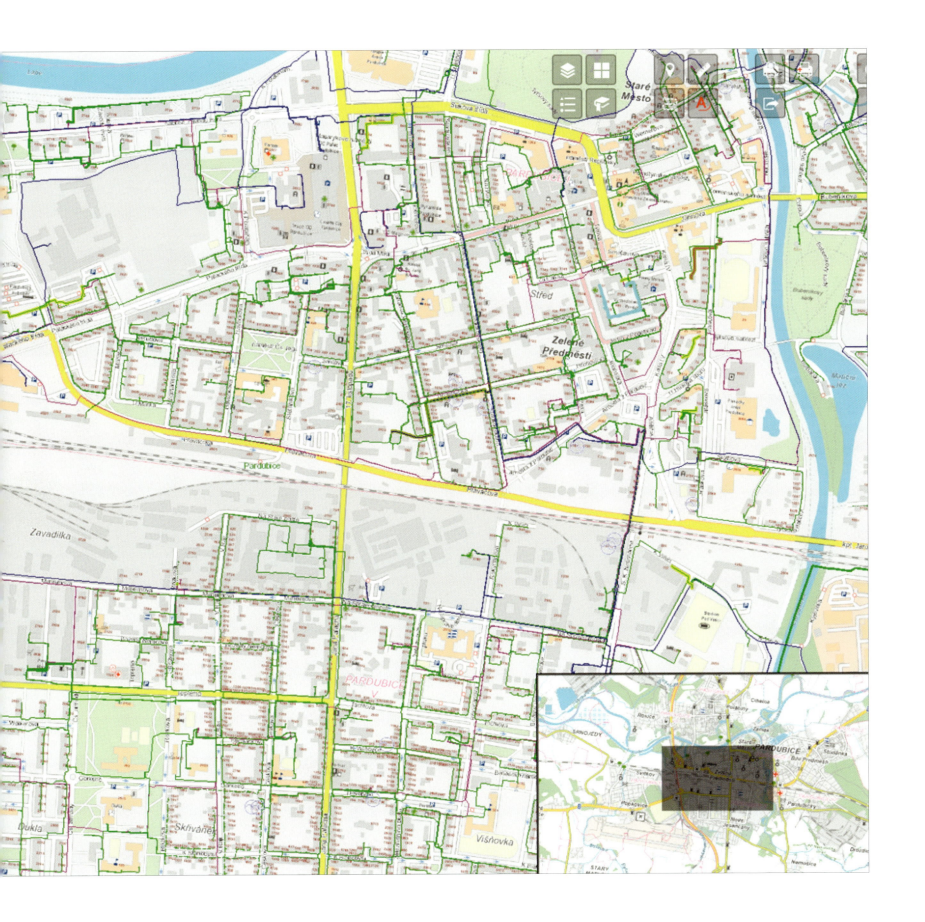

PREDICTING IMPACTS OF STORM INUNDATION TO SEWER STRUCTURES USING PYTHON

Manatee County Government
Bradenton, Florida, USA
By Ashley Snyder

Major storm events, such as hurricanes, require accurate maps produced efficiently. During Hurricane Irma, a map was requested of the sanitary sewer network structures that could be impacted by the potential storm surge flooding. The process for creating the map from scratch was time consuming, and the request went unfulfilled because new data became available before the map was completed.

An efficient process was required for the 2018 hurricane season. The result was a map template and accompanying Python code. Once the potential storm surge flooding data is downloaded from the National Hurricane Center (NHC), the Python code can produce a map within five minutes. The code was tested using the NHC data produced during Hurricane Irma.

The code first clips the NHC data to the county boundary, converts it from raster to vector, and applies the NHC storm inundation symbology. Next, the sewer network structures are clipped to the storm surge data, and the appropriate symbology is applied. A series of SelectByAttribute functions are performed to get a count of the total number of affected structures and of each structure type by owner. The map title is updated to the current storm name and advisory number. Lastly, the map is exported to a .png file that can be added to the situation report.

CONTACT

Ashley Snyder
ashley.snyder@mymanatee.org

SOFTWARE

ArcGIS Desktop 10.5.1

DATA SOURCES

Manatee County Utilities Records and Locate Division, NHC

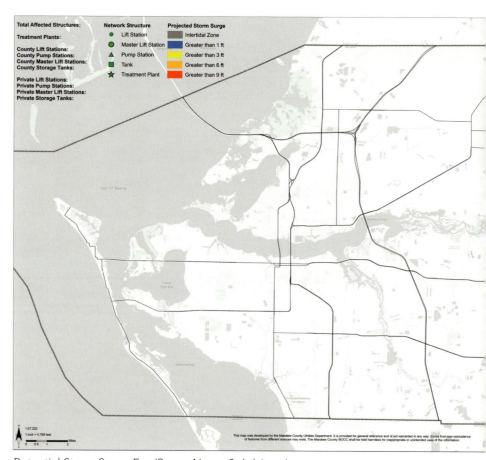

Potential Storm Surge For (Storm Name & Advisory)

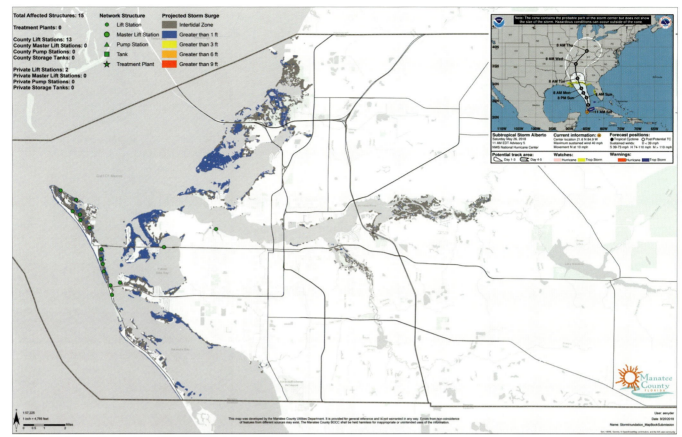

Potential Storm Surge For ALBERTO Advisory 05

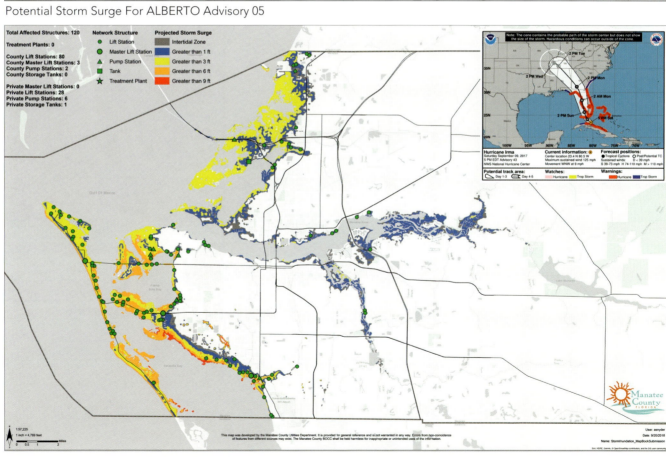

Potential Storm Surge For IRMA Advisory 43

COST OF SERVICE: URBAN DEVELOPMENT AND STORMWATER MANAGEMENT

Philadelphia Water Department
Philadelphia, Pennsylvania, USA
By Henry Bernberg

The Philadelphia Water Department (PWD) is the water, sewer, and stormwater utility for the city of Philadelphia. PWD calculates stormwater management service fees for nonresidential customers based on a parcel's size and impervious area while residential parcels are assessed a flat rate charge. The original impervious area data was captured in 2004 and updated data was collected in 2015, during which time the city has seen significant land development.

This map shows the percentage of a parcel's total area that changed from pervious to impervious between 2004 and 2015 and vice versa. Philadelphia saw a total increase of 8.5 percent in parceled impervious surface area, with approximately 66 percent of that resulting from new residential development. Due to PWD's rate structure, only 10 percent of nonresidential customers may see a change in their stormwater charge based on the updated data. Parcels that see a significant increase in their impervious area can be targeted for the implementation of stormwater management practices, which reduce impact on Philadelphia's stormwater infrastructure.

CONTACT
Henry Bernberg
henry.bernberg@phila.gov

SOFTWARE
ArcGIS Desktop 10.3.1

DATA SOURCES
PWD, City of Philadelphia

Courtesy of Philadelphia Water Department.

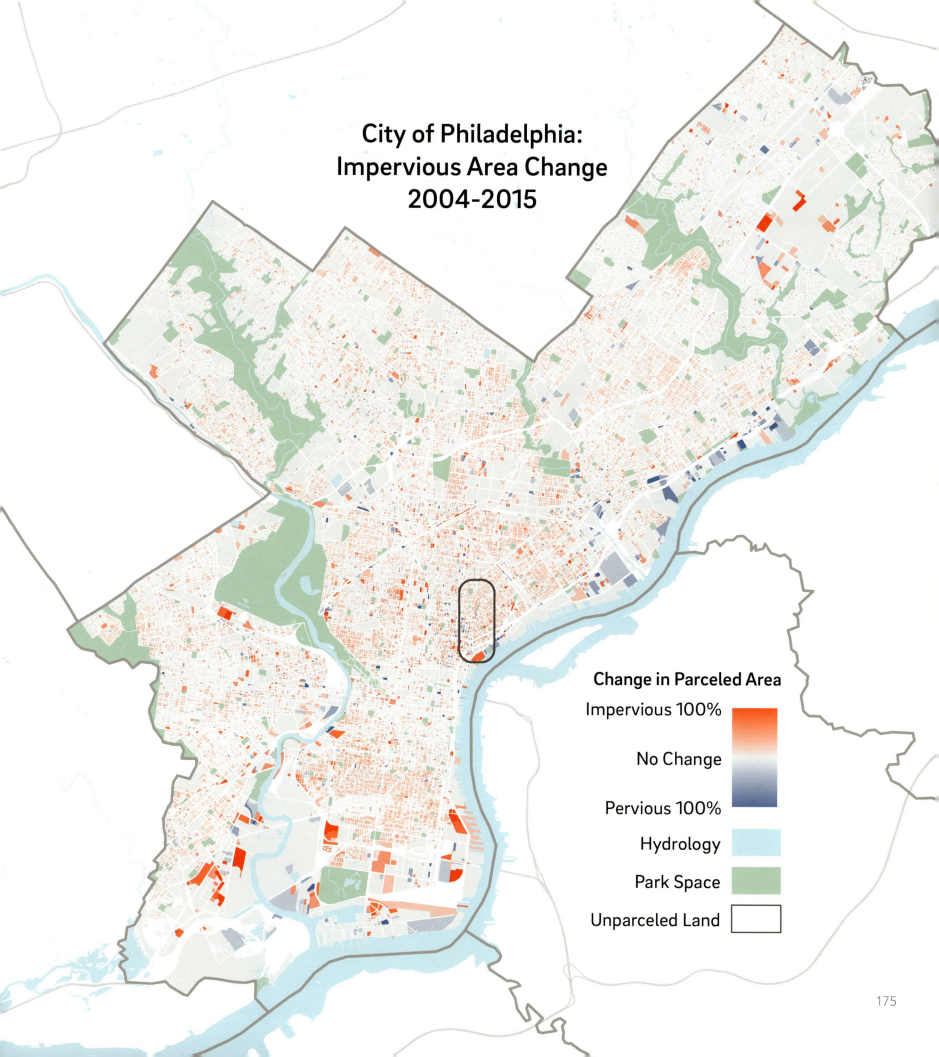

City of Philadelphia:
Impervious Area Change
2004-2015

Change in Parceled Area

Impervious 100%

No Change

Pervious 100%

Hydrology

Park Space

Unparceled Land

DOE SPATIAL WATER DEMAND FORECASTING SOLUTION

Department of Energy (DOE), Abu Dhabi, United Arab Emirates
By Shaima Al Hammadi

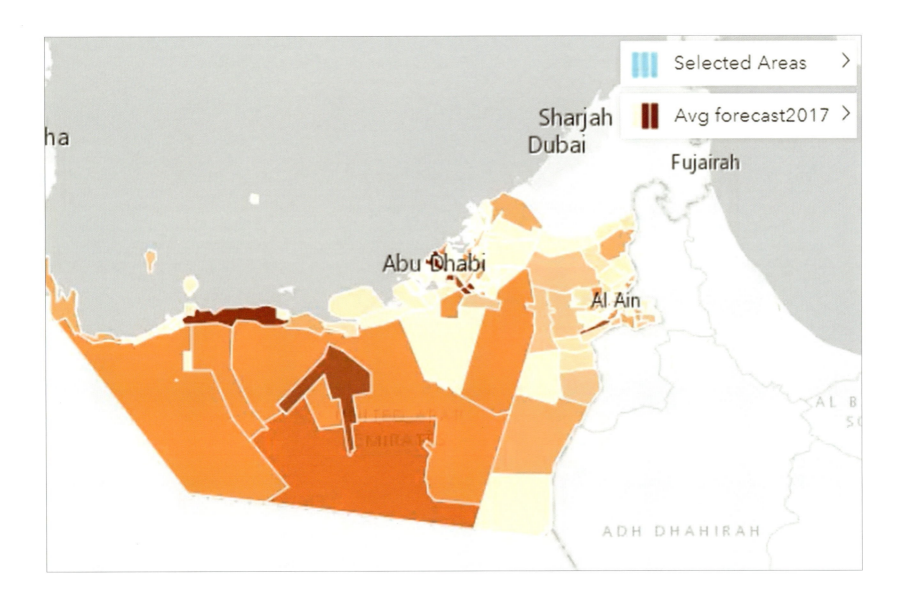

The Department of Energy (DOE) spatial water demand forecasting solution utilizes Insights℠ for ArcGIS® to better enable planners and forecasters to understand the spatial and temporal patterns of water demand and population trends. This provides them with the analytics needed to help optimize system operations.

With these maps the planners and forecasters can quickly detect patterns of population change between different years. They can also access customized cards that display the breakdown of the population in a specific year between citizens and expatriates, as well as the amount of housing by type and the trend

in population growth in the coming years. Similarly, in other tabs, the water demand data is plotted against the population data to observe the intrinsic patterns and correlation between demand, population, and time, especially with DOE implementing sustainable initiatives to reduce water and power consumption and demand.

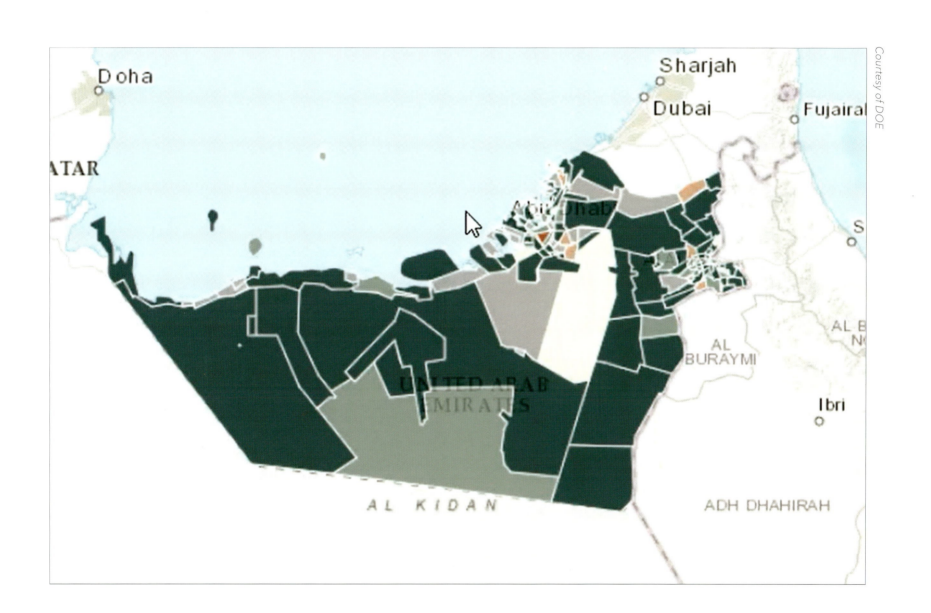

CONTACT
Shaima Al Hammadi
shhammadi@addc.ae

SOFTWARE
Insights for ArcGIS

DATA SOURCES
DOE/Abu Dhabi Water and Electricity Company